D0999680

Glorious Eclipses

Their Past, Present and Future

This beautiful volume deals with eclipses of all kinds – lunar, solar and even those elsewhere in the Solar System. Bringing together in one place all aspects of eclipses, and lavishly illustrated throughout, *Glorious Eclipses* covers the history of eclipses from ancient times, the celestial mechanics involved, their observation and scientific interest. Personal accounts are given of recent eclipses, up to and including the last total eclipse of the twentieth century: the one on August 11th 1999 that passed across Europe, Romania, Turkey and India. This unique book contains the best photographs taken all along its path and is the perfect souvenir for all those who tried or wished to see it. In addition, it contains all you need to know about forthcoming eclipses up to 2060, complete with NASA maps and data, making it the perfect resource for both novice and veteran eclipse-chasers.

Serge Brunier
Jean-Pierre Luminet

Glorious Eclipses

Their Past Present and Future

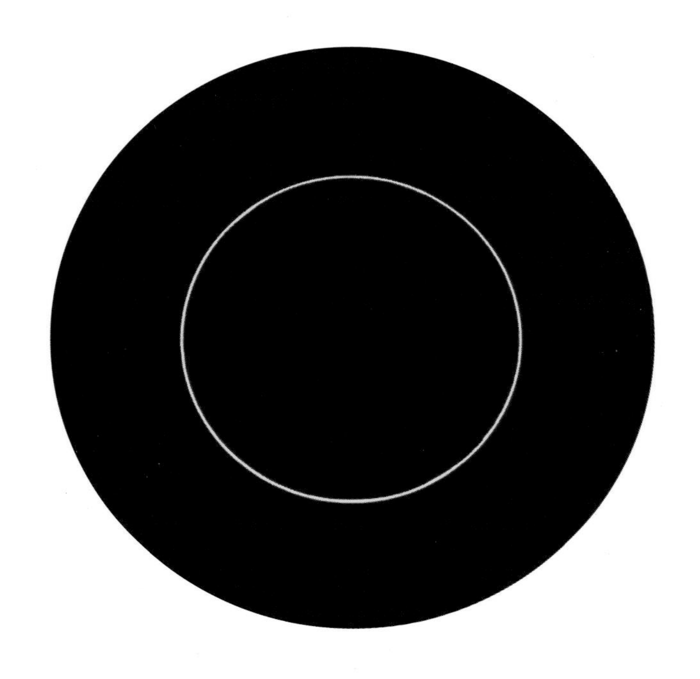

Translated by Storm Dunlop

BORDAS

PUBLISHED BY THE PRESS SYNDICATE OF THE UNIVERSITY OF CAMBRIDGE
The Pitt Building, Trumpington Street, Cambridge, United Kingdom

CAMBRIDGE UNIVERSITY PRESS
The Edinburgh Building, Cambridge CB2 2RU, UK http:/ /www.cup.cam.ac.uk
40 West 20th Street, New York NY 10011-4211, USA http:/ /www.cup.org
10 Stamford Road, Oakleigh, Melbourne 3166, Australia
Ruiz de Alarcón 13, 28014 Madrid, Spain

First published in French as *Eclipses, les rendez-vous célestes*, by Serge Brunier & Jean-Pierre Luminet 1999
First English publication 2000

Printed in France by POLLINA, Luçon, France - n° L 80686

Typeface Zapf Humanist 10/14.5pt. *System* QuarkXpress® [HM]

A catalogue record for this book is available from the British Library

ISBN 0 521 79148 0 hardback

Contents

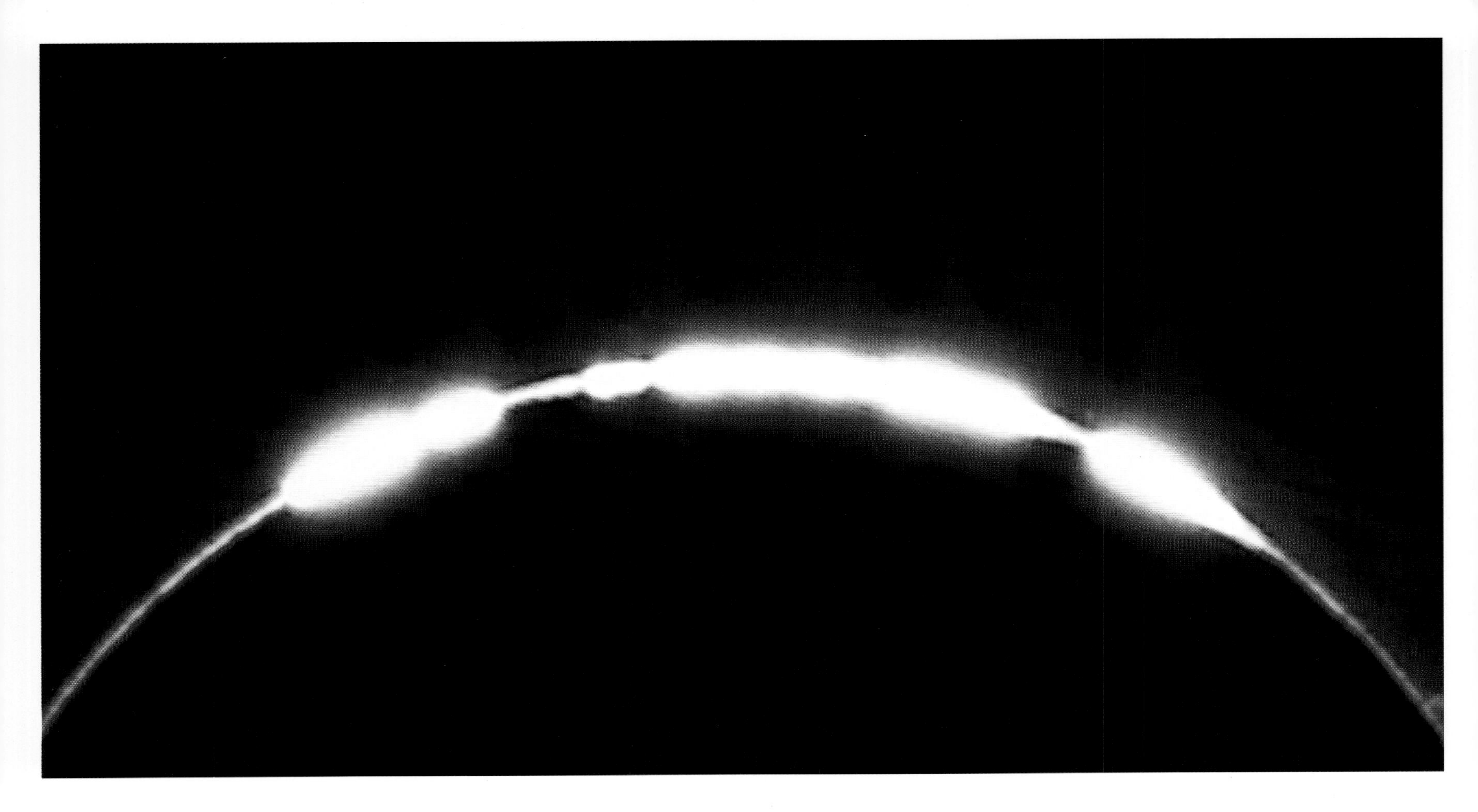

Henceforth there will be two types of Europeans: those who saw the eclipse of 11 August 1999, and those who were unable, or simply did not wish, to admire the great heavenly encounter. This book has been conceived for both. For the former, we hope that these pages will perpetuate the memory of the most beautiful sight offered by nature. As for the latter, I, and Jean-Pierre Luminet, would like to convince them to go and see, with their own eyes, the next total eclipse of the Sun. And although it is true that we would have to wait until 12 August 2026 to see another total eclipse in Europe, and even until 3 September 2081 before the Moon's shadow crossed France from one side to the other, in reality total solar eclipses occur nearly once a year somewhere on Earth. The last eclipse of the millennium was also the seventy-seventh in the last century!

The next encounters between the Sun and Moon are already fixed by the laws of celestial mechanics: 21 June 2001, 4 December 2002, 23 November 2003, and 8 April 2005 are the first few of the new millennium, but it will not always be easy to travel to the distant countries or deserted regions where eclipse chasers will be: Madagascar, Mozambique, the Antarctic, and the South Atlantic...

Passionately interested in astronomy ever since the age of twelve, for me eclipses remained, for a long time, simple dates in the ephemerides, and I had to wait until I was thirty-three before witnessing, for professional reasons, my first total eclipse, that of 11 July 1991, from the Hawaiian Observatory on top of Mauna Kea volcano.

It would be an understatement to say that I immediately became passionate about the celestial events, which I have followed ever since, over the course of the years and the lunations, more or less all over the planet. Each time, there is the same astonishment and, each time, the feeling has grown that eclipses are not just astronomical events, that they are more than that, and that the emotion, the real internal upheaval, that they produce – a mixture of respect and also empathy with nature – far exceeds the purely aesthetic shock to one's system.

On reflection, it is perhaps because what we are celebrating in eclipses is a secret collusion that intimately links mankind with the Sun and Moon. In the last few decades, with the conquest – through our robot spaceprobes – of the Solar System, astronomers have come to realise the strangeness of the Earth–Moon pair, which is unique among the nine planets and sixty-odd satellites that orbit our star. Although the relationship between the Sun and life is immediately and instinctively apparent, it is only very recently that the idea has arisen that perhaps the appearance of life on Earth, and its subsequent evolution as far as *Homo sapiens sapiens,* may have been possible solely thanks to the presence of the Moon within the blue planet's sphere of influence. Over the eras and aeons, the Moon would have stabilized and protected the Earth and its inhabitants, a bit like the way a balancing pole stops a tightrope walker from falling into space.

And where do eclipses fit in all that? Everyone who has admired one, will tell you that they strike a distant, yet profound, note in the very depths of your soul.

I like to think that a few million years ago, when in the savannas of southern Africa, early humanoids were slowly learning to walk upright, their first representatives were given a push towards conscious thought by observing a total eclipse of the Sun. Night that fell in the daytime; stars at midday; the blinding brightness of the Sun vanishing, to be replaced by an incomprehensible and terrifying well of shadow; the strange indescribable beauty of the diaphanous corona surrounding this dark abyss that had suddenly appeared in the dome of the sky, and then the miraculous return of light... A total eclipse of the Sun: a striking and dramatic encapsulation of the beauty and mystery of the world.

Serge Brunier
Paris

Why should a theoretical astrophysicist, who has rarely peered through the eyepiece of a telescope, preferring to speculate on the invisible architecture of space-time by means of dry equations, be interested in eclipses, to the extent of writing a book about them?

It is true that Einstein's general theory of relativity, which has been a constant intellectual delight for me over so many years, was first proved experimentally thanks to a 'simple' solar eclipse. That was on 29 May 1919. But such an argument satisfies only the intellect. When it comes to the soul, it demands a greater spectacle, a more tangible emotion.

I was not yet ten on 15 February 1961, when a total eclipse of the Sun crossed the Provence where I was born. All I can recall is preparing smoked glass under the watchful eye of our schoolmistress. Cloud over Cavaillon probably spoilt the spectacle, because I have no memory of the eclipse itself...

Then... then I waited nearly forty years before finally seeing a total eclipse of the Sun. That was on 26 February 1998, in the Sierra Nevada de Santa Maria, in the north of Colombia. Serge Brunier and a few other colleagues, genuine eclipse chasers, had finally convinced me that any astrophysicist worthy of the name should not 'die a fool'! It is true that, obviously, such a recurrent and 'nearby' astronomical event might seem somewhat prosiac to someone who spends his life in abstract research on models of the Big Bang, black holes, and the realms of space-time.

On that day, however, between 10:59 and 11:03, I experienced what John Couch Adams described so well, some 150 years before. The English mathematician-astronomer, less accustomed to handle telescopes than complex equations of celestial mechanics – he predicted the existence of the planet Neptune through calculation – witnessed a total eclipse of the Sun for the very first time in his career during the summer of 1846. He subsequently described the extraordinary emotions he felt as an astronomer – and, moreover, an experienced one – who realized that he was a novice when he discovered this astounding cosmic drama for the very first time.

The darkness that descends suddenly in the very middle of the brilliance of the day; this new light that arises from the obscurity; the planets aligned like a necklace of pearls in a configuration that is never seen by night; and the Sun's flamboyant corona...

These few minutes in which time seems suspended, create the almost palpable feeling of being, transiently, part of the invisible harmony that rules the universe. It is as if a sudden opening in the opaque veil of space allows our inner vision to reach into the otherwise hidden depths of the cosmos, giving us humans – mere insignificant specks of dust – an all-too-brief instant to see the other side of the picture.

To me, the invisible is not restricted to dark objects that our telescopes cannot detect. It is also, and in particular, the secret architecture of the universe, the insubstantial framework of our theoretical constructs. I have always been moved by black. Not black as in absence, but rather black as revealing light. According to the painter François Jacqmin 'Shadow is an insatiable star-studded watchfulness. It is the black diamond that the soul perceives when the infinite rises to the surface.' Astrophysics and cosmology team with examples where black is all-important. It is in the black of the night that one sees stars, or, in other words, that one perceives the immensity of the cosmos. This same night-time darkness reveals the whole evolution of the cosmos, and the finite nature of time. The mass of the universe is largely dominated by dark matter; massive, non-luminous objects, which through their gravitational attraction govern the dynamics of the cosmos. As for black holes, the epitome of invisibility, they are perhaps secret doorways opening onto other regions of space-time.

Another aspect of eclipses enthrals me: their historical and cultural dimension. I have always been attracted by the way in which different forms of human invention interact. Science, despite being an effective and rational approach to truth, nevertheless remains incomplete. Art, philosophy, and the comparative study of traditions, myths and religions, are all complementary approaches that are indispensable for anyone who wants to gain a greater insight into where they fit in the cosmic scheme of things.

In unfolding the story of people, their civilizations, and their relationships with heavenly phenomena, one cannot but be fascinated to see how past eclipses have influenced their course. The impression of a supernatural power engendered by the sudden disappearance of the Sun or the Moon has often struck human beings, frightened by an apparently hostile and incomprehensible nature, to the extent of changing their behaviour.

That's enough. My colleague, Serge Brunier, and I decided to write a book that alternated between these 'two voices'. After all, the various types of eclipses are created by the Earth and the Moon taking it in turn to pass in front of each other beneath the blazing Sun...

Eclipses are a benign contagious virus, which once it has infected you, recurs at intervals. Which is why, for the eclipse of 11 August 1999, I went deep into the Iranian desert, to be (almost) certain of finding a sky devoid of clouds.

Jean-Pierre Luminet
Meudon

Journeys of an eclipse chaser

■ On 11 July 1991, at 7:28 in the morning, a total eclipse of the Sun plunged Mauna Kea Observatory into darkness. The summit of the volcano, which is the site of some of the most powerful telescopes in the world, hardly emerged from the sea of clouds that covered the island of Hawaii. The line of the horizon lay outside the Moon's shadow, which crossed the Pacific Ocean at a speed of 9000 km/h.

■ THE SUN RISES SOMEWHERE OVER THE ATACAMA DESERT IN CHILE. IT IS ACCOMPANIED BY THE MOON, WHICH IS STILL INVISIBLE. THE ECLIPSE WILL START IN A FEW HOURS' TIME...

It always begins the same way. Sitting on the terrace of a cafe, the photographer watches resplendent, multi-coloured butterflies fluttering overhead: disturbing creatures who are lazily gliding in a deep purple sky that is slowly becoming darker and darker. With a glass in his hand, he watches this strange night that is falling, and the Moon that is slowly swallowing the Sun. He would like to move, to go down onto the beach of reddish sand, where all his powerful optical equipment is set up to record the eclipse, but he is unable to tear himself away from the table. When he finally does get up, and tries to make his way through a dense crowd of indifferent people to his telescope, a deep blue night has fallen, so beautiful that it takes his breath away, and he realises, as he runs between high towering buildings that deliberately hide the darkened Sun, that it is too late. He will miss the eclipse yet again...

This nightmare, and infinite variations upon it, is one that regularly comes to haunt eclipse chasers during the long months of waiting before the 'next one'. Because it is an extremely strange astronomical calendar that governs the lives of these men and women. Their existence is punctuated by meetings that are of absolutely vital importance, but which they can neither describe nor explain, and of which the principal characteristic is that they cannot be deferred. They know, two, five, or ten years in advance, precisely where – and at exactly what time – they will be on the planet.

But whatever are they searching for, six or ten thousand kilometres from home, on the coast of the South Atlantic, swept with rain, in the early hours of the morning, or in the middle of an unknown steppe in Mongolia, in a very uncertain spring?

To them, total eclipses of the Sun are not just celestial events. They are more than that. They link the Earth with the Universe; they demonstrate the underlying harmony of the world; and they tell us about the laws of space and time. To be there, at the right place and at the right time, is to assert a particular bond with the universe. There is, for example, at the very beginning when the edge of the Moon begins, almost imperceptibly, to bite into the blinding disk of the Sun, a secret and perhaps slightly infantile jubilation at such a well-ordered event; this absolute adherence to the astronomers' calculations. There is also that sudden, heart-stopping sense of cosmic perspective when the Sun finally covers the Sun, and abruptly, in a vast planetarium that is the size of the heavens, three bodies are absolutely precisely aligned. And finally, and above all, there is the most beautiful natural spectacle.

During the last hundred years of this second millennium, there have been seventy-seven total eclipses of the Sun. The first, 18 May 1901, plunged the south of Madagascar into

■ One of the most beautiful eclipses of the century took place in the Andes on 3 November 1994. That morning, the Moon's shadow crossed the Altiplano and the highest volcanoes in Bolivia and Chile. This photograph was taken on the shores of Lake Chungara, at an altitude of 4600 m.

■ Beyond the photosphere – the blinding surface of the Sun – and the chromosphere, the solar corona extends for millions of kilometres. It is a searingly hot, but highly rarified gas, seen here, as on the opposite page, at the eclipse of 11 July 1991.

darkness at dawn, then the islands of Réunion and Mauritius, before crossing the Indian Ocean, the islands of Sumatra, Borneo, and New Guinea, and ending at sunset in the Pacific. The last total solar eclipse of the millennium took place over Europe and Asia on 11 August 1999, before the eyes of several tens of millions of people: probably a historical record. Each one of these eclipses was a completely unique event. Each total eclipse had its own playhouse, its own drama, scenery, sudden developments, and finally, its own audience, who gave it life and transformed it into history.

Seventy-seven total eclipses – that seems a lot; almost one a year. Yet less than 0.5 per cent of the Earth's surface is involved each time. Because 70 per cent of our planet is covered with water, 20 per cent of its land is basically desert, and because cloud cover over the inhabited regions adds a touch of uncertainty to the impeccable celestial ballet, the sight of a total eclipse is, for most people, exceptionally rare. The chances of seeing one – without deliberately travelling elsewhere to see one, of course – is essentially zero: just once every four hundred years. Who can remember the eclipse of 21 September 1903, visible only from the latitude of the Furious Fifties, in the Antarctic Ocean? Who recalls the one in 1904? At dusk on 9 September, after having crossed the southern Pacific Ocean, the shadow of the Moon finally disappeared in the sands of the Atacama Desert, at the foot of the Cordillera of the Andes in Chile. Was there anyone that evening, between the oases of Copiapo and Antofagasta, to appreciate the night that fell *before* the Sun set?

■ At the observatory on Hawaii, on 10 July 1991, several astronomers watch the very last sunset before the great heavenly encounter the following morning.

The last decade of the millennium was to offer eclipse-chasers some memorable encounters with the Sun and the Moon. Above all, their view of the most beautiful of Nature's spectacles was to change. Until then, scientists has appropriated eclipses and, with them, a crowd of amateur astronomers who rushed around the world to help them or, at least, imitate them. A total eclipse became synonymous with a stopwatch, a tape recorder, telescopes, and radial filters. To them, the phenomenon was a scientific one, and involved only those who were familiar with the sky. The aim was to measure, to record, to calculate, and to confirm

■ When the Moon's disk precisely hides that of the Sun, the outer atmosphere of our star appears. Here, two prominences, more than 100 thousand km in height, extend from the delicately coloured, pink chromosphere.

the event in detail. All too frequently the terrestrial element of the event was forgotten, and with it, unfortunately, the magic of that much-anticipated moment... The only important thing was to obtain a picture of the black disk of the Moon against the silvery background of the corona.

That was a great shame. All the more so, because more than a century ago, Camille Flammarion, a poet, astronomer, enlightened visionary, journalist, and writer, was one of the first, in his immense work *Astronomie Populaire* (*Popular Astronomy*), which appeared in 1880, to enchant his readers with his lyrical depictions of total eclipses, described by him in a happy choice of words as a spectacle that had never been seen before, and would never be seen again. In his book, he finishes his account of total eclipses of the Sun by giving the reader a list of all partial, total, or annular eclipses that would be visible up until 1900. Alas! Who among his readers would have been able to travel on 7 July 1880 to Cape Horn, on 12 December 1890 to Mauritius, or on 29 September 1894 to Madagascar, to enjoy a few moments of intimacy with the cosmos? An eternal optimist, Camille Flammarion nevertheless wrote: 'Unfortunately, not one eclipse of the Sun will be total in France; but so long as our inventions for the use of steam and electricity continue and other new ones come along, it will not be long before the Earth will be but a single country, and we will be able to travel from here to Peking with less fuss than it took a century ago to journey from Paris to St Cloud.'

■ On the eve of this eclipse, the twilight displayed unbelievable colours, caused by volcanic dust hurled into the sky by the eruption of Pinatubo several weeks earlier.

Most eclipse-chasers nowadays – including the writer – discovered their all-consuming interest as a child or adolescent, in dry yearbooks of astronomical ephemerides, calculated up to the year 2000. These were often difficult for the uninitiated to decipher, but which, with the help of a little imagination, promised a veritable feast of events to come. Although the year 2000 seemed in the remote, almost incomprehensible future, the list of dates held a greater fascination; it was associated with magical, inaccessible places. To take just the last decade: 1991, total eclipse, North Pacific, Central America; 1992 total eclipse, South Atlantic; 1994, total eclipse, Peru, Chile, Bolivia, Brazil; 1995 total eclipse, India, South-East Asia; 1997, total eclipse Mongolia, Siberia; 1998, total eclipse, Central

America, Caribbean; 1999, last total eclipse of the millennium, Europe.

Camille Flammarion's prediction, has, of course, been fulfilled. In the 1980s, for the first time, men and women – although not very many of them, it is true – were able to proudly boast: 'I have seen fifteen total eclipses of the Sun.' Strangely, although all of them marvelled at the heavenly alignment that caused the Sun, up there, to become dark, none of them really took much note of their earthly surroundings in which they had experienced those precious moments. With hindsight, the reason became clear. Their whole outlook and their thoughts were exclusively astronomical. If, on average, the narrow band in which the eclipse was visible was some 14 000 km long, i.e., about one third of the Earth's circumference, they systematically chose the site where the total phase of the eclipse lasted as long as possible. The 'great' eclipse of 1991 was to give some of them the desire and the opportunity to rediscover the sense of occasion that Camille Flammarion expressed. The date of 11 July 1991 was in the thoughts of all eclipse specialists, both amateur and professional.

That day, in fact, thanks to a particularly favourable combination between our planet, our star, and our satellite, the total phase of the eclipse promised to last for 6 minutes 53 seconds – a real piece of eternity. The only thing was that, to benefit from this exceptionally long period of time, one needed to travel to Mexico, where the Sun was eclipsed more or less at midday, and very close to the zenith. Thousands of enthusiasts from all over the world, from the nearby United States, to as far away as Japan, as well as from the Old World of Europe, made just that choice. They went to see a magnificent eclipse, of extraordinary duration, right above their heads, but which was lost in an immense, blue sky. Magnificent actors, but without any true theatre in which to put on their performance.

That year the eclipse began in the middle of the North Pacific Ocean. Before crossing Central America from one side to the other, and then ending at sunset in the Amazon basin, it would fleetingly pass over the island of Hawaii. It is there, not far from the Tropic of Cancer, that the volcano of Mauna Kea lies, rising 4,208,m above the waves of the Pacific. At the top of the volcano, the largest astronomical observatory in the world has been built. The majority of the most powerful telescopes in the world have been gathered together up there, in their gleaming domes, some of which as large as the *duomo* of the cathedral of Santa Maria del Fiore in Florence.

■ At 7:48 on the morning of 11 July 1991, the last ray of light from the Sun escapes between two lunar mountains. In less than a second, the total eclipse will begin over Mauna Kea Observatory.

An island above the clouds

In Hawaii, the duration of the eclipse would not exceed 4 minutes 28 seconds. Taking place in the early morning, the encounter between the Sun and the Moon would occur due east, at just 21° above the horizon. To astronomers, such conditions are not at all favourable. All the more so, because although there was a great temptation to use the giant instruments at the site for observing the eclipse, they had not really been designed to observe the Sun. There was, in fact, a considerable risk that the mirrors of the observatory's telescopes, each measuring some metres across and designed to capture the light from stars hundreds of millions of times fainter than the faintest stars visible to the naked eye, might damage the telescopes, by accidentally concentrating the light from our own star. Despite this very real danger, numerous international teams were unable to resist the urge to direct the telescopes and radio telescopes on Mauna Kea – including one of the most powerful instruments of the time, the 3.6 m Canada-France-Hawaii Telescope – towards the eclipsed Sun (*see* Chapter 6). Some photographers chose this site, a wonderful vantage point for viewing the universe, over the beaches of Baja California, which would be packed with thousands of astronomers, attracted by one of the longest total eclipses of the century. The evening before the eclipse, all the protagonists met, shivering, at the summit to admire the sunset. All sensed, without being able to see it, the presence of the Moon not far from the star that was sinking behind the horizon. The two bodies were due to meet a few hours later.

It was just like that persistent nightmare: on the morning of 11 July, at the moment when the Moon's shadow plunged Hawaii into darkness, it was pouring with rain on the slopes of Mauna Kea. At an altitude of 3000 m, where the several dozen scientists and photographers accredited by the University of Hawaii had a restless sleep, many had the same bad dream. Much lower down, the road had been closed by the rangers, who feared that the mountain would be

■ Along the crest of Mauna Kea volcano, at an altitude of more than 4200 m, the domes of the Hawaiian observatory are silhouetted against the twilit sky. High above, the Moon will hide the Sun for precisely 4 minutes 28 seconds.

■ Early morning on the shores of Lake Chungara, on the border between Bolivia and Chile, the volcanoes Pomerape and Parinacota seem to await the forthcoming eclipse. That day, the 3 November 1994, the Moon had already started to cover the Sun, but to the naked eye, their movements were still imperceptible.

invaded by a horde of people, more accustomed to golden beaches than to the rigors of high altitudes. But the scientists who woke at 3 o'clock in the morning to snatch a quick cup of poor coffee and eggs 'sunny side up', realised that the nightmare was only too real: Mauna Kea was well and truly buried in the clouds, and a warm, persistent, tropical rain was soaking the ancient lava flows, once fluid themselves, but now solidified for thousands of years.

It was 4 o'clock. From Hale Pohaku, the base camp, built at an altitude of 3000 m, the 4 x 4s started to slowly claw their way up the 20-km slope that leads to the observatory. The rain seemed likely to last for ever, so there would be no eclipse. Exhausted by their bad night and by the weeks of preparation, the astronomers were gloomy, and nearly dozed off again, watching the monotonous movement of the windscreen-wipers. At 3800 m, the vehicles were spattered with wet ashes, the windscreens were covered, and the headlights had difficulty in penetrating the darkness: it was a moonless night, a starless night, a hopeless night in a wasteland of black ashes... Now it was 3900 m. The scientists and the few rare journalists invited along for the eclipse were sunk in the depths of depression.

And then, suddenly, the 4 x 4s passed the 4000-m barrier and broke out into a clear black sky, above the clouds that were blanketing Hawaii. The summit of Mauna Kea was like an island above the clouds. It was a miracle: up there it was fine, and in the growing dawn, the last stars were fading in the sky. At 6 o'clock, the Sun had just risen, shedding an orange-violet glow across the sky. In their domes, the specialists were busy: preoccupied with their computers, overwhelmed by the amount of work to be completed and by the delicate and dangerous task of manipulating their instruments, they were oblivious to the most magnificent sight in Nature. Outside, there was a science-fiction atmosphere. Several photographers has set themselves up, a few tens of metres apart, on one of Mauna Kea's secondary cones. They wanted, of course, to obtain the one photograph that no one had ever taken, and which no one would ever take again for several centuries; the ideal image, the archetypal image of astronomy: the sight of a total eclipse above an observatory. One of them, Barney, readjusted an oxygen mask on his face: this was his first eclipse, and his heart was racing fit to burst, and he had not acclimatized to the altitude.

Time comes to a halt on Mauna Kea

It was 7:27. Fingers trembled over shutter releases. The Moon had almost completely obscured the disk of the Sun, and the light had become dusk. On the mountain, where the thermometer had fallen to below 0°C, everyone was fighting the cold. The hair-thin crescent of the Sun cast strange shadows across the arid landscape. The starless sky was abnormally dark. With wildly beating hearts, and in breathless silence, the few spectators see, towards the western horizon, the shadow of the Moon blotting out the sea of clouds as it races towards them at an unnatural rate. Everything happens at the same time, and much too quickly. The light suddenly disappears, the observatory domes are silhouetted against a blue-black night sky, the disk of the Sun crackles with points of light – just as it is hidden behind the mountains at the lunar limb – and then, time comes to a halt. Dumbfounded by the incredible novelty of the sight – never before seen, as Flammarion said – the photographers hesitate before releasing their shutters.

■ HIGH UP, AT AN ALTITUDE OF 6342 M, IN THE FREEZING COLD, AND UNDER A CRYSTAL-CLEAR SKY, A FEW ASTRONOMERS ADMIRED THE ECLIPSE FROM THE SUMMIT OF PARINACOTA VOLCANO. ON EITHER SIDE OF THE SUN, WHICH WAS SURROUNDED BY A VAST CORONA, THE PLANETS JUPITER AND VENUS APPEARED.

The sight is so staggering, so ethereal, and so enchanting that tears come to everyone's eyes. It is not really night. A soft twilight bathes the Mauna Kea volcano. Along the ridge, the silvery domes, like the ghostly silhouettes of a temple to the heavens, stand rigidly beneath the Moon. The solar corona, which spreads its diaphanous silken veil around the dark pit that is the Moon, glows with an other-worldly light. It is a perfect moment. We ought to be able... But it is too late, the eclipse is over.

High above, the Sun has suddenly reappeared. The ochre, grey, and black volcanic cones slowly emerge from the darkness. The white domes close to cracking and low groaning sounds; and all that remains is the normal, peaceful, morning scene at the Mauna Kea Observatory. Everyone, awe-struck and in silence, prepares to go back down the mountain. But for some of the spectators, infected by the bug, an insidious, insistent question recurs: when is the 'next one'?

In its own way, the next eclipse was destined to enter the annals of astronomy. It was on the verge of the southern winter, the 30 June 1992, on the coast of Uruguay. At the deserted whaling station of Punta del Este, lashed by an appalling drizzle, sweeping in off the Atlantic, a few half-frozen amateur astronomers awaited the dawn at the edge of the sea. But the winter had already decided otherwise, and so that morning no one saw the sight that no one has ever described, and of which there is therefore no trace in any historical archive: a total eclipse that occurs precisely at the time of sunrise...

So after the total eclipse of 11 July 1991, eclipse chasers had to be patient for more than three years, until the next one, which awaited them in the early hours of the morning of 3 November 1994. For 1211 days, or 42 lunations, the Sun and the Moon were to pass one another every month, without meeting, or at least only at arm's length. With annular eclipses and partial eclipses, and its perfect precision, the great celestial chronometer marked the passage of the seasons.

SNOW AND GOLD ON THE TWIN VOLCANOES

Faced with a blinding slope of snow and ice, an exhausting climb towards a sky that grew ever darker, the Sun rising on the eastern horizon in a freezing cold dawn, a wind that seemed to be driven by malevolent desire to harm and which made every step towards the summit seem like yet another station of the cross, the small group of astronomers, lost on the slopes of Parinacota volcano, saw their dream of eclipses turn into a nightmare. This eclipse was one of the most beautiful of the century, and there were many enthusiasts who would not miss it for anything. On that morning, 3 November 1994, the shadow of the Moon was to cross South America from one side to the other, plunging one of the most beautiful unspoilt natural wildernesses on our blue planet into darkness. On the Andean Altiplano, on the borders of Peru, Bolivia, and Chile, the perfect cones of two volcanoes that are among the highest in the world, Parinacota and Pomerape, are reflected in the calm waters of Lake Chungara. It is difficult to imagine anywhere that is at the same time so harsh and so beautiful.

Whilst the group of climbers exhausted themselves trying to arrive at the summit of Parinacota before the start of the eclipse, a few freezing photographers, just out of their tents, prepared to admire Nature's most wonderful spectacle from the rather more reasonable altitude of the lake: 4600 m. The night had not been an easy one for them either. The day before they had to find the

■ During the eclipse of 3 November 1994, the planet Venus was to be found close to the Sun and Moon in the sky. Such a close approach between the brightest of the planets and the eclipsed Sun will not occur again until 2 August 2027.

ideal spot to set up their telephoto lenses opposite the lake, the volcanoes and, of course, the forthcoming eclipse. They had hours of bumping along in 4 x 4s, under the rather aloof stare of the vicuñas and, above all, under an ominous-looking sky. Would the eclipse really occur tomorrow? And then night fell, but without any comforting twinkling of stars. A dark sky, overcast, and hopelessly uniform. Then there was the cold, the silence, and the oppressive darkness. Towards midnight, snow began to fall, and the water of the lake started to freeze. The pink flamingos, standing endlessly on a single leg, weren't worried. But on the shores of the lake, the observers started to get in a panic, seeking any pretext to leave the endless solitude of the Altiplano. Going back down: hit by the beginnings of acute mountain sickness, which they had not really taken seriously before they found themselves nauseated, breathless, their heads aching with a dreadful migraine, two amateur astronomers from Europe could think of nothing else. Their willpower was being insidiously eroded by hypoxia, fatigue, and the cold, and in the end the eclipse did not seem very important. A lot of people, in the middle of the night, started to make their way down, to the great astonishment of the Chilean customs officers, who had expected an increasing flood of vehicles in the opposite direction on the road to Bolivia and the Tambo Quemado pass, which marks the frontier between the two countries, at an altitude of 4800 m.

Peter, the stolid German eclipse chaser, who thought he had seen it all before in the harsh Bavarian winter, gave in at midnight, and with his friend Frans turned back on the road to Putre, the comforting and welcoming village that lay more than 1000 metres lower, and where thousands of tourists who were frightened of the Andes were sticking together. By the time early morning came to the shores of the lagoons, between the twin volcanoes, and opposite the forthcoming eclipse, would there be nothing but playful viscachas – mountain chinchillas – graceful vicuñas, and indifferent flamingos?

In the freezing dawn, however, a few stars timidly appeared in a waxen, but calm, sky. After that interminable night, which had not allowed anyone to sleep, the Sun finally rose into a sky of royal blue, still scattered with a few high-altitude clouds.

Just before dawn, Peter and Frans returned to the Altiplano, bucked up by the 'recompression' by more than one hundred millibars that they had gained at Putre, which lay at an altitude of 'only' 3500 m. Looking at the Parinacota glacier through a small astronomical telescope, Peter was amazed to see a column of climbers working towards the summit, after a bivouac at an altitude of more than 6000,m! They would also see the eclipse, but what state would they be in? On the shore, a few flamingos were warming themselves in the first rays of the Sun. The eclipse could begin. At 8:27, the Moon obeyed the astronomers' calculations and started to nibble away at the Sun. Although, when observed through a dense filter, the progression of our satellite across the disk of the Sun is very striking, nothing is visible to the naked eye. Our eyesight cannot cope with the brilliance of the Sun. It remains blinding, even when its surface – and thus its brightness – is reduced by a factor of ten, one hundred, or even a thousand.

Menacing clouds invade the sky

A few minutes before totality, when the Sun is still as bright, and when, unfortunately a few menacing cirrus clouds reappear in the slowly darkening sky, Peter notices that Venus is perfectly visible to the naked eye, just 5° from the Sun! At about 9:17, hearts start to pound, and breath – already speeded up by the altitude and emotion – becomes short. The panorama presented by the cordillera is absolutely astounding. Towards the west, the sky darkens in just a few seconds, turning to midnight blue. Suddenly, a dark shadow rushes across the landscape, and the final blinding point of light disappears, where the Sun used to be. In this other-worldly night, Venus, Jupiter and Mercury blaze with a steady light, whilst the solar corona spreads out in vast silvery wings. With numb and trembling fingers, shocked by the supernatural beauty of the spectacle, the photographers fire off their cameras without thinking.

What they feel, in that high-altitude desert, cannot be shared. No other experience can compare. It is like some primitive race memory, a bridge across the centuries and millennia to all the other civilizations that had that experience – a mixture of sympathy and compassion for the ancients who were terrorized by this supernatural phenomenon – and also pride and exhilaration at the idea that, nowadays, it is not the wrath of a hostile and incomprehensible nature that causes eclipses, but the precisely ordered motion of a

■ Eclipse chasers wait years for their encounters with the Sun and the Moon. But the sight lasts only a few minutes; once one is over, the wait begins again.

■ On 24 October 1995, nearly ten thousand people gathered at the ghost city of Fatehpur Sikri, built in the 16th century by the Grand Moghul. They were

small celestial body, on which a few of mankind's representatives have actually set foot.

In the distance, towards Peru and Bolivia, where the Moon's shadow has yet to arrive, the Andean peaks glow faintly, while the snows on Parinacotta are tinged with gold. The Moon and the Sun seem to hang fixed to the dome of the sky for all eternity. But at the lunar limb the delicate pink arc of chromosphere appears at the bottom of a crater; a sign that in a second we will have to come back to earth, the eclipse will be over.

A shadow play at Fatehpur Sikri

It was just two or three kilometres short. Christine had looked at the problem from every angle, poring over her map of India yet again, and running over in her mind the magnificent sites that would be covered by the shadow of the Moon on 24 October 1995. There was no way out. Nature, this time, had decided to play a cruel trick on eclipse fanatics. No one would see a weird night fall over the magnificent marble mausoleum that Shah Jahan had built in 1631 for his dead wife, Mumtaz Mahal. The thousands of people who would congregate on the banks of the serene Jumna that reflects the four flawless minarets of the Taj Mahal, would see 99.8% of the Sun hidden, but no total eclipse. As for the famous observatories constructed by the maharajah astronomer Jai Singh at Delhi and Jaipur, specifically to predict eclipses, at the appropriate time they would see a piece taken out of the Sun by the Moon, but their gigantic sundials would not be plunged into darkness.

From her maps and atlases, Christine finally discovered where she should go to admire what would remain one of the most spectacular eclipses of the century. One day, more than four hundred years ago, the Moghul emperor Akbar decided to build his capital and palace in the middle of the desert. Fatehpur Sikri, a magnificent city of red sandstone, bristling with ramparts, minarets, and high towers was built in a few years, and remained the Moghul capital for fifteen, and then was

ABLE TO ADMIRE AN EXTREMELY IMPRESSIVE ECLIPSE. HERE, THE TOTAL PHASE LASTED LESS THAN FORTY SECONDS.

suddenly deserted in 1586. Here, in this ghost city, on the 24 October 1995, along with Christine, nearly ten thousand other people gathered, from every state in India, and from all over the planet. It is perhaps a way of showing the heavens that the works of mankind also make their mark on the passage of time.

That morning, around 8 o'clock, the Sun slowly stops fighting against the Moon, and is inexorably extinguished. The palaces of Fatehpur Sikri lose their colour, becoming fantastic dark silhouettes against an indigo sky. Suddenly, from the terraces of Panch Mahal a resounding cry goes up. Hundreds of people see an incredible sight – one never seen before, as Camille Flammarion said. Racing across the rich plains of Rajasthan at more than 2255 m/s, the shadow of the Moon is about to engulf the city of Fatehpur Sikri. Almost without realizing it, Christine fires her camera, as she is fascinated by the beauty of the eclipse. The limb of the Moon is ringed with pink, and a dozen tiny flames lick out into the lower corona. For an instant, whilst still blindly firing the shutter, she stares up at the sky. She has just time to see, above the darkened walls, a black disk in a midnight-blue sky, a dark, disquieting eye, rimmed by pink, and surrounded by a silvery glory. A truly cosmic spectacle, both stupefying and utterly bewitching, that leaves everyone who witnessed it overcome with emotion. An image that was also too fleeting, that needs to be replayed later, so that it will be properly remembered.

Because hardly had the eclipse begun, than the first rays of sunlight reappeared between the mountains on the lunar limb. At Fatehpur Sikri, situated on the edge of the band of totality, the eclipse lasted less than forty seconds. It was just this short duration, which implied that the disks of the Sun and the Moon were almost identical in size, that gave this eclipse its particular magic. By just hiding the Sun's disk, the Moon left the very edge of the Sun uncovered, with its delicate pink chromosphere and prominences. Bemused, Christine looks at the film counter on her camera with astonishment, and wonders if everything worked properly. A little distance away, Sanjay Sharma, an Indian friend, watches the rejoicing crowd. The young Hindu is more impressed than he might at first appear by the event that he has just witnessed. He remembers that in the Mahabharata Epic, the fundamental text of Hindu lore, the great battle of Kurukshetra, in which Vishnu, the supreme god, was revealed, took place on a day when there was a total eclipse of the Sun.

A COLD CLEAR LIGHT ON THE LANDSCAPE

It always begins the same way. At the top of a dangerously swaying tower, a thousand metres high, the astronomer tries desperately to set up his equipment. The wind is blowing in gusts and, every time he tries to line up his fat, red plastic telescope, it overturns. In the deep purple sky, the total eclipse started long ago. On both sides of the darkened Sun, several crescent Moons, lined up as if on parade, add to the amazing beauty of the scene. While, up there in the sky, the eclipse lasts for hours, and the plastic telescope slowly deflates in a corner, the astronomer checks a giant wall calendar to see whether he has got the wrong date. When it finally seems that everything is ready, he realises that down on the horizon, the Sun and its attendant Moons is about to set: he will miss the eclipse yet again...

For once, on the morning of 26 February 1998, in the Paraguana peninsula in Venezuela, everything was going too well. Patrick, the French photographer, already had three totals to his credit and had, moreover, not yet failed with any. He wondered if today his luck would change. He was in Hawaii in 1991, Chile in 1994, in India in 1995 and, each time, everything went wonderfully well. Of the two total eclipses in the last decade that he had decided not to try to see, the one in 1992 in Uruguay, and the one in 1997 in Mongolia, no one else had seen anything. In both cases, clouds hid the sky for eclipse chasers. But here, in the coastal desert fringing the Caribbean, the atmosphere before the eclipse is astonishingly calm. The wind, which generally blows for weeks on end, fell at 12:30, just as the Moon started to nibble away at the Sun. It was a marvellous start to the afternoon; the overwhelming midday heat of the tropics had given way to a gentle warmth; the blue of the sky deepened and, in the contorted divi-divi trees, sculptured by the prevailing winds, the confused humming-birds had gone to roost. Two o'clock came. The Sun is still high in the sky, not far from the zenith, still blindingly bright, but the landscape is illuminated by a soft, cold light, a sort of unchanging twilight, giving a supernatural air to the cactus desert that was normally baking in the heat. In the abnormally blue sky, above the cactus plants, which were casting weaker and weaker shadows, Patrick noticed the planet Venus. No wind, not a cloud, and the photographer is finally ready. He knows now that he will see the eclipse. The thought makes him dizzy. This is perhaps the most intoxicating moment of an eclipse, the instant when, after years of waiting, you find that you are at exactly the right time, at the right place, with a fine sky overhead, and with a strange sense that something magnificent and preordained is going to happen. Up there in the sky, a cosmic ballet of unimaginable dimensions is getting ready and overtaking you, and nothing can stop you from witnessing it: the total eclipse is about to take place. It is 14:08. Now, everything will happen too fast. Lost among the cactus, which have been reduced to ghostly silhouettes, Patrick finally risks looking directly at the Sun. For an instant, the universe appears to hang in the balance. It is broad daylight, the Sun still remains blinding, but seems to be disintegrating. Then, suddenly, it as if someone up there turned out the light. The whole sky darkened. The Sun had disappeared, replaced by a dark disk, fringed with a radiant, silvery aura, shot through with pink flecks of light: a diaphanous veil of such incredible beauty that it took your breath away. In the strange twilight, with its cold, metallic light, the photographer sees that the eclipsed Sun is watched over by two other brilliant bodies: Mercury and Jupiter.

■ THE PLUMES OF THE CORONA – A GASEOUS ENVELOPE A MILLIONTH OF THE BRIGHTNESS OF THE DISK OF THE SUN ITSELF – ARE VISIBLE ONLY DURING TOTAL ECLIPSES, AS HERE, ON 26 FEBRUARY 1998.

Patrick doesn't take his eyes away, all the time operating his cameras for form's sake – he knows that no photograph can ever reproduce the beauty of this sight, that you have to see it to understand it – and he tries in vain to make the most of this unexpectedly intense moment. It is now 14:11 and 45 seconds. At the lunar limb, a blinding flash announces the end of the eclipse. Once again, this incredible dream has passed all too quickly. Yet again, Patrick, bemused, wonders what had happened, and why the sight is so beautiful and why it affects his very soul. And as he will not be able to find an answer, and as he already feels the loss, like everyone else, he will start to count the days until the 'next one'. ■

■ On 26 February 1998, a strange twilight fell over the desert on the Paraguana peninsula, in Venezuela. The soft, metallic light of the solar corona – roughly comparable with that of the Full Moon – bathed the landscape for 3 minutes 20 seconds. Mercury and Jupiter stand guard over the eclipsed Sun.

The story of eclipses, the story of people

■ In this magnificent pastel drawing made by E.L. Trouvelot of the total eclipse of the Sun on 29 May 1878, observed from Wyoming, the solar corona, visible for just a few brief minutes, presents a strange appearance. The coronal streamers around the poles, which seem like fountains of light, and the broad equatorial plumes, like two giant wings, have given rise to various legends in a whole range of cultures.

■ IN ANCIENT CHINA, SOLAR ECLIPSES WERE CLASSED AS 'FRIGHTFUL' AND 'ABNORMAL'. THEY WERE THOUGHT TO BE CAUSED BY A DRAGON THAT TRIED TO DEVOUR THE DAYTIME STAR.

The term eclipse comes from the Greek word *ekleipsis*, which means 'abandon', 'failure'. If one did not know that the disappearance of the Sun in broad daylight during a solar eclipse is caused by the temporary intervention of the Moon, would one not fear that light had abandoned the world, that this sudden night would last for ever? Would one not imagine that it was a sign of divine anger, or the work of some malevolent power of darkness? In the same way, when a lunar eclipse occurs, the way in which the Full Moon is engulfed in a vast area of shadow, and the way its colour changes from white to blood-red certainly indicate that there is some failure in the cosmic order.

For the majority of ancient peoples, the eternally repeating cycle of heavenly motions was a constant comfort. By momentarily breaking this harmony, an eclipse naturally engendered fear.

COSMIC DISORDER OR DIVINE PRESENCE?

In early Chinese, the ideogram shih, used to designate an eclipse, and which is found scratched on certain animal bones, means 'eat'. The Chinese did, in fact, believe that eclipses were caused when an invisible dragon tried to devour the Sun or the Moon. To scare away the monster, the whole population was ready to create as much noise as possible, with the frantic ringing of bells, beating of gongs and drums, while archers fired arrows into the sky. Even in the 19th century, the Chinese imperial navy fired their ceremonial guns during eclipses.

The idea of a ferocious animal that is trying to swallow the heavenly luminaries is common to many civilizations. In India, people immersed themselves up to the neck in the Ganges; Voltaire described the scene: 'It has not yet proved possible to abolish the ancient Indian custom of immersing oneself in a river at the time of an eclipse; and although there were Indian astronomers who knew how to calculate eclipses, the general populace remained convinced that the Sun was being swallowed up by a dragon, and that it was not possible to release it without throwing oneself naked into the water, and by making a lot of noise to frighten the dragon into letting go of its prey.'

In Central America, the Maya were convinced that during an eclipse of the Moon, it was being eaten by a giant jaguar. In the *Dresden Codex*, a Mayan calendar for prophesying the future, the symbol for an eclipse hangs above the open mouth of a serpent.

In North America, the Chippewa Indians, fearing that the fires of the Sun would be utterly extinguished by an eclipse, used to

Le Petit Journal

ADMINISTRATION
61, RUE LAFAYETTE, 61
Les manuscrits ne sont pas rendus
On s'abonne sans frais
dans tous les bureaux de poste

5 CENT. SUPPLÉMENT ILLUSTRÉ 5 CENT.
23me Année — Numéro 1.118
DIMANCHE 21 AVRIL 1912

ABONNEMENTS

	SIX MOIS	UN AN
SEINE et SEINE-ET-OISE..	2 fr.	3 fr. 50
DÉPARTEMENTS..........	2 fr.	4 fr. »
ÉTRANGER	2 50	5 fr. »

Conserver ce numéro pour prendre part au Concours.

CEUX QUI VIRENT L'ÉCLIPSE DE 1724
CEUX QUI VEULENT VOIR CELLE DE 1912

■ In modern times, eclipses have lost their mystery. On 17 April 1912, an annular eclipse of the Sun still enthralled the crowds gathered at St Germain. The last eclipse visible at Paris was observed by the young king Louis XV in 1724.

■ Worshipping Tonatiuh, the solar god of the Aztecs. The large disk surrounded by red dots symbolizes the Mexican cycle of fifty-two years. Standing upright on a podium, beneath

fire burning arrows in its direction in the hope of relighting it.

Among the Tarahumara of Mexico there was a 'Black Sun' sacrificial rite, Tutuguri, described in a hallucinatory text by Antonin Artaud.

In Japan, wells were closed to prevent the sky poison, hidden by the eclipse, from falling into them.

Almost everywhere on the planet, the battle between the powers of light and the powers of darkness, or the voracious appetite of mythical animals were the basis of the explanations of eclipses.

A passage in *The Revelation of St John*, written towards the end of the 1st century, faithfully describes the sensations that may be felt during an annular eclipse of the Sun: 'Then I watched as he broke the sixth seal. And there was a violent earthquake; the Sun turned black as a funeral pall and the Moon was all red as blood.' Not being completely obscured by the

WHICH AN INVERTED CONE PROBABLY INDICATES AN EARTHQUAKE, TWO RICHLY CLOTHED DIVINITIES OBSERVE A TRIPLE DISK OF RED, YELLOW, AND WHITE, WHICH PROBABLY SYMBOLIZES A SOLAR ECLIPSE.

Moon in an annular eclipse, the lower atmosphere of the Sun, known as the 'chromosphere', remains visible, and the bright red prominences might appear like spurts of blood.

The text of *Revelation* establishes a link between eclipses and earthquakes. The idea was not new. In the 5th century BC, the Greek historian Thucydides, chronicling the Peloponnesian War, had already stated that in those days 'earthquakes and eclipses of the Sun occurred more frequently than in earlier times'. In addition, he recorded that 'there was an eclipse of the Sun at the time of the New Moon, and in the first part of the same month there was an earthquake'.

Another Greek historian, Phlegon, reported a similar event: 'In the fourth year of the 202nd Olympiad, there was an eclipse of the Sun that was greater than any other known previously, and at the sixth hour, the day became night. Stars appeared in the sky, and a

■ A conjugal eclipse, engraved by Granville in 1844. Some people had a romantic view of eclipses: the Sun and Moon are lovers who, when they embrace, turn out the light to preserve their intimacy. Here, a plethora of indiscreet telescopes spies on their behaviour.

■ In the 18th century, the Spanish explorer Don G. Juan described the Peruvians despair during an eclipse. Chanting, dancing, drums, and the howling of dogs that were being beaten were thought to chase away the malevolent beast that was devouring the Sun.

great earthquake that struck Bythnia destroyed most of Nicea.' The eclipse concerned was that of 24 November in the year 29, the possible date of the crucifixion of Jesus.

The desire to demonstrate a cause-and-effect relationship between these two types of phenomena, one celestial and the other terrestrial, doubtless reflected the attempt to derive a certain regularity for seismic events from the well-established cycle of eclipses. The tradition went back to the Babylonians; passionately interested in astrology, they saw celestial events as portents that governed terrestrial affairs. This type of confusion seems to persist today in certain regions of the globe. The earthquake of 16 September 1978 which ravaged Iran, the most devastating one that year, because it killed more than 25 000 people, occurred three-and-a-half hours before a total eclipse of the Moon was visible in the region. Many of the inhabitants did not consider the coincidence to be fortuitous.

Celestial lovers

Some races, however, were not frightened by eclipses. The Eskimo and the Aleuts interpreted them as a sign of good fortune; the Sun and the Moon temporarily left their natural places in the sky to reassure themselves that everything on Earth was going well. Some eclipse legends are love stories, rather than accounts of war. Among the Aborigines, the Moon and Sun, husband and wife respectively, pull the curtains in the sky to ensure privacy for their union. The Tlingits of British Columbia and the Tahitians, conversely, consider the Sun as male, and the Moon as female; the rest of the story is precisely the same...

It is not always easy, on an ethnological level, to be certain of the beliefs of races that are loath to communicate with Western culture. I was able to see this for myself at the time of the total eclipse of the Sun that took place on 26 February 1998 in Central America and in the Caribbean. The path of totality ran, in particular, across the Sierra Nevada de Santa Marta, a mountainous range in the north of Colombia. The south-eastern slopes of the mountain are inhabited by the Arhuaco Indians. Anxious to preserve their traditions, they avoid contact with the rest of the country. At Nabusimake, the main village in the reserve, our group of astronomers and anthropologists, consisting of Frenchmen and Colombians from Bogotá, tried to question the Indians about their beliefs. Mistrustful of strangers, they had little to say, limiting themselves to agreeing with the replies that we suggested. Are the Sun and Moon two lovers who embrace during an eclipse? Nods of the head and a glimmer of a smile. Was it because of prudery that the union of the two bodies should not be watched? Difficult to know. What remained certain was that according to ancient tradition, the Indians would shut themselves away in their huts at the moment of total eclipse. Each of them, men, women, and children, curiously accepted a strip of exposed photographic film that we had brought with us, and started to observe the eclipse during its partial phase with cries of astonishment. Then, as the dark disk of the Moon bit farther and farther into the bright disk of the Sun, one by one they left the square. At the moment of totality, all the Arhuacos had disappeared into their huts. Only a handful of westerners remained on the village square to admire the amazing sight.

■ This Latin manuscript, composed in the middle of the 11th century illustrates the sixth vision of Saint John, in which the Sun becomes as dark as sackcloth and the whole Moon becomes the colour of blood.

The prediction of eclipses as an instrument of power

Unlike comets, which, for a very long time, were considered as enigmatic, unpredictable events, eclipses were the subject of quite accurate predictions at a very early stage.

Tradition in the Far East shows us, in fact, that the ruling classes were able to predict the dates of eclipses around 2300 BC. Their predictions were based on an empirical law, governing the recurrence of events, by virtue of which the relative positions of the Earth, Sun, and Moon, recur in precisely the same manner after 6585 days (*see* Chapter 4).

■ Eclipse observation in China around 1840. Whilst the astronomers calmly observe the phenomenon with a telescope, the terrified servants prostrate themselves on the ground to placate the ill omen.

In Assyria, cuneiform tablets show that thirty-five centuries ago, this eclipse cycle, known as the Saros, was equally known to the Babylonians. The latter were able to calculate lunar eclipses in advance with an accuracy of a few minutes. Similarly, in Central America, the Mayan astronomers had discovered the principal celestial cycles, and their 12th-century calendars included the prediction of eclipses and their durations.

Although they were capable of predicting the dates of eclipses, these peoples did not, however, understand the mechanism. We will see in a later chapter that the very existence of a regular eclipse cycle such as the Saros is the result of remarkable coincidences, which involve complex combinations between the movements of the Moon, the Earth, and the Sun. Calculation of the narrow zone in which solar eclipses were visible was itself uncertain. Such knowledge, beyond the grasp of ancient peoples, only began to be acquired during the golden age of Greek astronomy, the second century BC. Although astronomical observations allowed the prediction of eclipses, very early on, they did not, for all that, assuage people's fears about the events. According to Plutarch, the fear engendered in earlier times by eclipses was explained by an ignorance of their causes.

■ In this *Account of a Jesuit voyage to Siam*, an eclipse observed in 1688 engenders interest among the scholars, and fear in the ignorant.

As civilizations evolved, the fact of being able to know in advance when eclipses would recur, naturally acquired an importance much greater than being able to predict the positions of the Sun and Moon, which were the basis of all calendars.

Among the peoples of classical Antiquity, the great mass of the ignorant and superstitious were always terrified by the sudden darkening of the two bodies that lit up the daytime and night-time skies. The interruption of the regular course of celestial phenomena gave rise to the fear of a fundamental discord in the natural order of things, of a revolt of obscure forces against the higher powers that controlled the world. Is not the term 'cosmos' synonymous with order and beauty?

An eclipse of the Moon or Sun that occurred during a battle has, on more than one occasion, caused a reversal of fortune for the side that was the least afraid, in fact of those whose commanders expected the event with the greatest confidence. The episodes of the siege of Syracuse or the capture of Constantinople, which we shall describe later, are examples. The influence of astrologers on the psychology of princes was therefore very great. In this respect, the similarities with comets are precise. The episode recorded on the Bayeux Tapestry, which describes the invasion of England by the Norman, Count William, in 1066, is well-known. Shortly

■ An imaginary and unrealistic representation of a ceremony at Stonehenge. The Celtic priests gathered people at this sacred site at dates prescribed by celestial phenomena. The symbol of the serpent, like that of the dragon, is associated with eclipses.

before the decisive Battle of Hastings, an immense comet (later to be called Halley's Comet) appeared in the sky. The Normans saw it as a favourable omen, whereas Harold's astrologers, in England, were frightened. We do not know if these opposite interpretations had any effect on the troops' morale, but the fact remains that the Normans gained the upper hand. Harold was killed, hit in the eye by an arrow, and William, henceforth known as 'The Conqueror', was crowned at Westminster on Christmas Day 1066.

It is interesting to realise that, among all the cultures that were capable of foreseeing these phenomena, predictions of eclipses were generally treated as doctrinal secrets, which were divulged to the general populace with a certain aura of secrecy. It was, in fact, a question of exerting the greatest possible influence over peoples' minds. This was possible only because there was, on the one hand, a scholarly elite that influenced those in power, and on the other, the illiterate masses, all of whom knew nothing of regular events in the heavens.

Bit by bit, the religious castes, and the state institutions, especially in China, came to understand the extent of the influence that they could derive from their astronomical knowledge. They introduced the practice of offering prayers and invocations at those times when an eclipse of the Moon or of the Sun could be expected. It was essential to prepare the populace; the approaching event was accompanied by rites designed to prevent any dire effects. At a stroke, the general fear was considerably diminished, if the eclipse occurred at the time predicted. The fact that prayers had been offered in time quieted peoples' minds, and the ability to predict celestial anomalies enabled governments to pretend to have an influence over the obscure powers involved.

Eclipses of the Sun, the prediction of which for a specific place was rather more arbitrary, saw the same preparations. If it did not take place, the effect produced on the populace was even greater, because it proved that the measures taken had prevented the danger. In fact, it was not uncommon for the total eclipse of the Sun to have cast its shadow on a neighbouring country, as would be learned later, in which case it was perfectly obvious that the priests' predictions and the government's decrees had fully protected the country. In short, the prediction of eclipses was used, for many years, as an extremely effective instrument of power!

The Stonehenge serpents

At Stonehenge, in southern England, there is an incredible group of enormous worked stones, or megaliths. This site has given rise to innumerable legends, poems, and scientific studies.

Why were these stones erected on a seemingly deserted spot on Salisbury Plain? Everything points to the fact that Stonehenge was a remarkably conceived astronomical observatory, perfectly adapted to its site. Every year, on 21 or 22 June, that is, on the first day of summer, known as the 'solstice', the Sun rises at its most northerly point in the whole year. At Stonehenge, an observer standing in the centre of the stones, would see the Sun rise at the solstice on the axis of the Avenue and pass above a large dressed stone, called the 'Heel Stone'. The change in the direction of the Earth's axis of rotation, caused by precession of the equinoxes, complicates any exact reconstruction of the alignments that may have been observed more than three thousand years ago.

But it is difficult to believe that this is pure chance, and the choice of the siting of Stonehenge was therefore of fundamental importance. Its builders probably wanted to commemorate the summer solstice – which is also the longest day of the year – as the beginning of the new year. By counting the number of days between two successive alignments, they could determine the length of the year, subdivisions of which would serve as a practical calendar for marking the seasons and feast days, time for sowing, for harvest, etc.

According to the archaeologists, the construction of the monument took place in three phases, lasting more than fifteen centuries, between 3000 and 1500 BC. It was during the second phase, around 1900 BC that Stonehenge was altered towards a specifically astronomical use.

Stonehenge has greatly excited people's imaginations. In the 1960s, some astronomers, such as Gerald Hawkins from the U.S.A., and Fred Hoyle from Britain, advanced the hypothesis that Stonehenge acted as a true 'computer', programmed to calculate eclipses! What is the truth of this?

Apart from the length of the solar year, to predict eclipses one needs to know two additional cycles. One, the length of the lunar month (a lunation) is easily determined. It is simply the number of days that separate two successive Full Moons. This cycle of 29 and a half days appears to be represented in Stonehenge's architecture by two circles, consisting of 29 and 30 holes respectively (known as the Aubrey Holes), the arithmetic mean of which is indeed 29.5.

The other cycle is far more complex: it is the period of rotation of two invisible points in space, known as the 'lunar nodes'. These are the points at which the orbit of the Moon, which is slightly inclined to the plane of the Earth's orbit (the ecliptic), intersect the latter plane. The Moon returns to one of these points every 27.2 days. But the line joining the nodes does not remain fixed in space: each node completes one revolution around the ecliptic in 6798 days (18.6 years), moving in a retrograde direction. This cycle, known as the 'lunar nutation period', governs the occurrence of eclipses: an eclipse may occur only when the Sun is more or less perfectly aligned with a lunar node (see Chapter 4).

It would seem that the builders of Stonehenge may have discovered this cycle and may have used it to predict eclipses. It doubtless required centuries of careful observation of the rising and setting Moon to determine the period of rotation of the nodes. This information, transmitted from generation to generation, could be coded into the architecture of Stonehenge. All the alignments necessary to determine the eclipse cycle are indicated by 19 megaliths placed in a horseshoe arrangement.

■ At Stonehenge, the position of the sunrise at the summer solstice every year is marked by a megalithic alignment, which gives credence to the view that the site was designed as an astronomical observatory.

If this hypothesis is correct, we can imagine the way in which the monument was used to keep track of the lunar cycles. The number of dressed stones or of pits dug in the ground around the different circles at Stonehenge, represented, for each of the circles, a certain number of days or years in the cycles. By moving markers (such as stones) along a circle in step with the cycles, it would be possible to follow the positions of the Sun, Moon and the two lunar nodes.

Periods when there was a probability of an eclipse occurring – that is, when the positions of the Sun and Moon were close to one of the nodes – could thus be predicted. If a Full Moon occurred at one of these periods, then all the conditions were present for a lunar eclipse to be visible from Stonehenge. It would be the same for a New Moon and the corresponding solar eclipse. Total solar eclipses at a given place are, of course, rare. But the law of averages means that about once a year – weather permitting, of course – either a partial solar eclipse, or a lunar eclipse may be seen from any given point on the Earth's surface.

Who were these stone-age people capable of observing such subtle astronomical cycles? The idea that Stonehenge could have been a religious site constructed by the Celtic priests, has often been advanced. However, the first Celts did not become established in Britain until around the 8th century BC. The megaliths thus date back to a far earlier epoch: the neolithic. Neolithic peoples doubtless had their own priests, in the form of shamans. It is tempting to imagine gatherings at a sacred place and at a time prescribed by the shamans (such as the solstices, equinoxes, and eclipses) to reaffirm their religous beliefs. What rôle might eclipses have played in these forgotten religions? Were they interpreted, as almost everywhere else, in terms of an unfavourable celestial omen, or did the shamans have a higher vision of the cosmic order?

Eclipses fully reflect the regularity of the major cosmic cycles. As for the invisible lunar nodes, they could have been integrated into the shamanistic religion in the form of *invisible gods*, capable of eclipsing the brightest bodies in the sky. The early British investigator of Celtic remains, William Stukeley, who was the first, in 1740, to note the solstice alignment with the Heel Stone, put forward the theory that Stonehenge and other similar megalithic sites were temples dedicated to the serpent god. How is the symbol of the serpent linked to eclipses? Let us recall that the key to determining

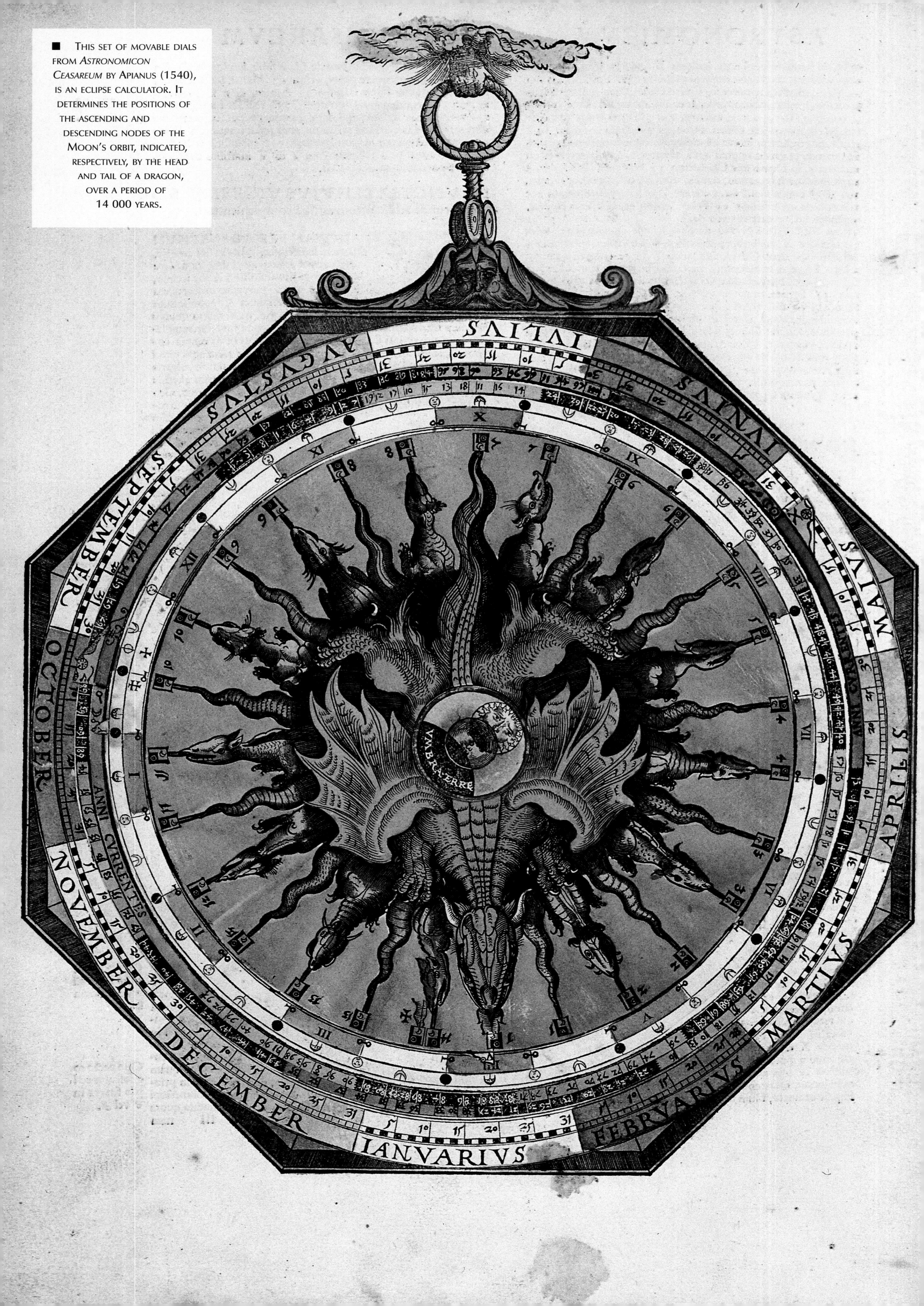

■ This set of movable dials from *Astronomicon Ceasareum* by Apianus (1540), is an eclipse calculator. It determines the positions of the ascending and descending nodes of the Moon's orbit, indicated, respectively, by the head and tail of a dragon, over a period of 14 000 years.

eclipses is the position of the lunar nodes. The period of time that the Moon takes to return to one of its nodes is called the 'draconitic month' by astronomers. In Latin, *draco* means both 'serpent' ('snake') or 'dragon'. The mythical serpent of Stonehenge and the legendary Chinese dragon are, perhaps, symbols of one and the same belief that originate in the depths of time: that of an invisible presence in space and time, which returns periodically to eclipse the Sun and the Moon.

■ IN THIS PAINTING ON A WOODEN PANEL FROM ANCIENT EGYPT, A PRIEST OFFERS INCENSE TO THE GOD RE-HORAKHTY. AT THE TOP, THE BODY OF THE SUN IS CARRIED BY THE TWO GREAT WINGS OF THE GOD HORUS. SOME HISTORIANS INTERPRET THIS SYMBOL AS REPRESENTING A TOTAL ECLIPSE WHEN SOLAR ACTIVITY WAS AT A MINIMUM.

THE WINGS OF THE SUN

While the Babylonians and the Celts were developing astronomical observation, the ancient Egyptian civilization flourished. The pyramids, temples and tombs are witness to the high degree of development reached by their art and technology. The Egyptians obviously were also involved in astronomy; the clear nights in the valley of the Nile being ideally suited for observation of the heavens. They measured the length of the year by watching the rising of Sirius, the brightest star in the sky. The largest pyramid at Gizeh is orientated towards the four cardinal points, and one of the inner passages is aligned on the star that was then at the North Pole, Alpha Draconis, the principal star in the constellation of Draco (the Dragon).

In 1369,BC, Amenophis IV, a pharaoh of the 18th dynasty, took the name of Akhenaton, and founded the cult of Aton-Ra, the Sun God. This monotheistic and reforming monarch, whose reign lasted seventeen years, celebrated the divine creative energy by a hymn to the Sun:

'Thy dawning is beautiful in the horizon of heaven, O living Aton, Beginning of life! When Thou risest in the eastern horizon of heaven,
Thou fillest every land with Thy beauty; For Thou art beautiful, great, glittering, high over the earth; Thy rays, they encompass the lands, even all Thou hast made.'

These verses are just a brief extract from a long poem in which the pharaoh makes no mention whatsoever of any possible darkening of the Sun by an eclipse.

To be honest, no representation of a solar or lunar eclipse is to be found at all evident in ancient Egyptian iconography. Yet a total eclipse of the Sun was visible from one part or another of the land of the pharaohs on average once every 75 years. It is difficult to believe that such a spectacular recurrence should not have been recorded in some form or other, especially among a people who were so versed in astronomy and who venerated the body that ruled the day. Numerous historians have investigated the question; some have suggested that the very particular appearance of the Sun during an eclipse might be represented in an allegorical form.

A valuable piece of evidence that allows us to confirm this idea is the Zodiac of Osiris. This circular vault in bas-relief representing a celestial planisphere was found in a temple at Dandarah (Dendera), in Egypt, during Napoleon's campaign. It is now in the Louvre Museum. The positions of the planets on the Dendera Zodiac allow us to give quite a precise date for the foundation of the temple, dedicated to Osiris, in which it was found. It was between June and August of the year 50 BC. Careful study of the cartouches reveals two symbolical representations of eclipses. An eclipse of the Sun is symbolized by a disk within which a goddess is pulling the tail of a baboon, which signifies that she it trying to liberate the Sun. The baboon is, among other things, a representation of the Moon in the form of the god Thoth. Another cartouche, which shows a disk enclosing a central eye, this time symbolizes an eclipse of the Moon. Astronomical reconstructions show that an eclipse of the Sun did occur at Dendera on 7 March 51 BC, in the region of the sky in which it is represented on the zodiac, that is, in the constellation of Pisces. In addition, a total eclipse of the Moon occurred on

■ IN THIS DRAWING BY SAMUEL LANGLEY OF THE TOTAL ECLIPSE IN 1878 AT SOLAR MINIMUM, THE EQUATORIAL STREAMERS RESEMBLE THE ANCIENT SYMBOL OF A WINGED SUN.

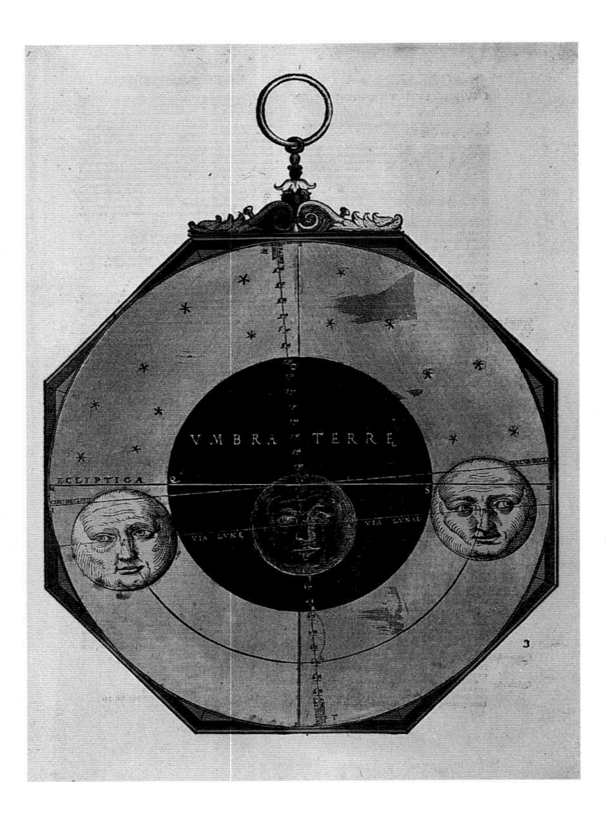

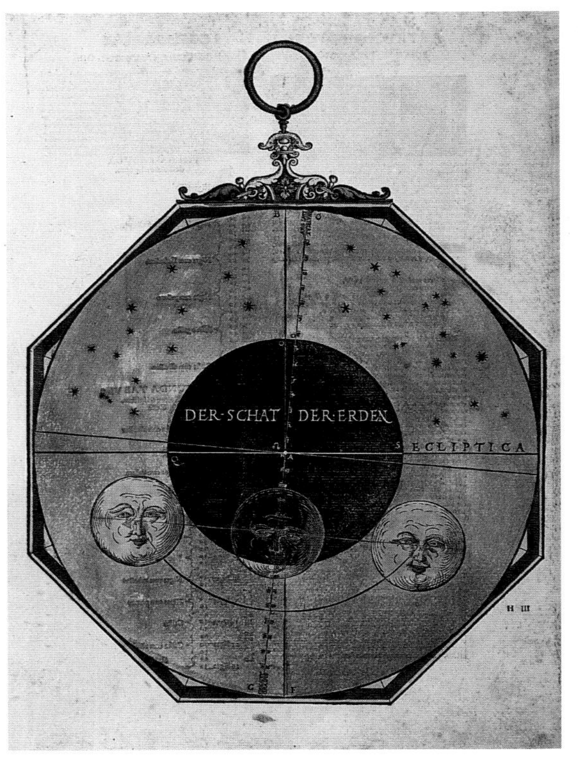

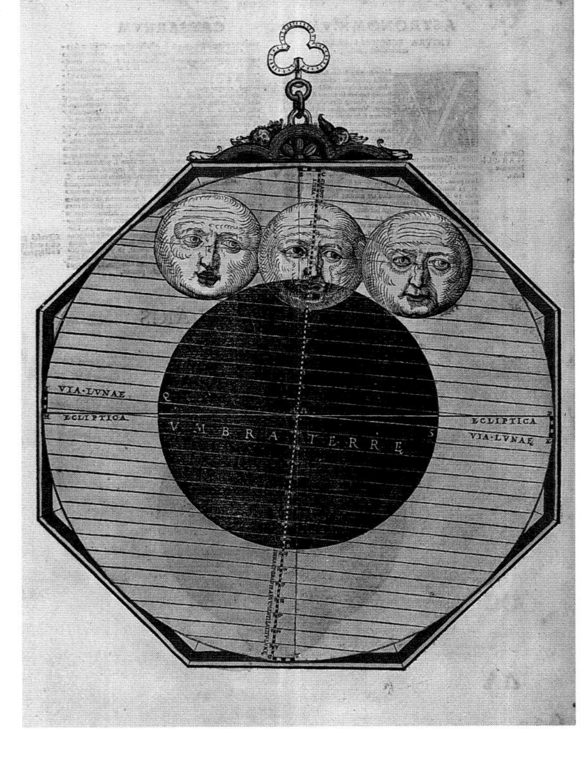

■ Eclipses of the Moon on 15 November 1500 (partial), 6 October 1530 (total), and 15 October 1502 (partial) as shown by Apianus in *Astronomicon Ceasareum*. The black disk represents the shadow of the Earth. Apianus noticed the variations in the colour of the Moon at a total eclipse.

24 September 52 BC, at the exact point where it is sown in the Dendera zodiac, in the constellation of Aries.

Another Egyptian symbol that may represent a total solar eclipse is that of the winged Sun – although this interpretation is controversial. This extremely old symbol figures over the entrance to numerous tombs and temples, as well as on papyruses. It is thought to represent the victory of light over darkness. Sometimes the design includes the heads of two serpents and the horns of a ram, which are also solar symbols. Modern astronomers have, moreover, retained the symbols and , which obviously resemble the horns of a ram, to indicate the ascending and descending lunar nodes. As we have already seen, these lunar nodes are closely associated with the occurrence of eclipses.

During certain total solar eclipses, the outer atmosphere of the Sun, normally invisible, appears in the highly characteristic form of a halo. The size and shape of this halo vary on an eleven-year cycle, corresponding to the sunspot cycle, which itself is linked to Sun's cycle of magnetic activity (see Chapter 6). At this cycle's minimum, the overall brightness of the corona is less intense, but long rays of light stretch out on either side of the Sun. Because these equatorial streamers are extremely faint, they are difficult to photograph, but with clear skies they are just visible to the naked eye during a total eclipse.

The similarity in shape between the solar equatorial streamers and the Egyptian symbol of a winged Sun is disturbing. Is the sight of a darkened, eclipsed Sun encoded in this symbol? The British astronomer E.W. Maunder, who studied the sunspot cycle stated the theory in these terms: 'there can be little doubt that the sun was regarded partly as a symbol, partly as a manifestation of the unseen, unapproachable Deity. Its light and heat, its power of calling into active exercise the mysterious forces of germination and ripening, the universality of its influence, all seemed the fit expressions of the yet greater powers which belonged to the Invisible.

What happened in a total solar eclipse? For a short time that which seemed so perfect a divine symbol was completely hidden. The light and heat, the two great forms of solar energy, were withdrawn, but something took their place. A mysterious light of mysterious form, unlike any other light, unlike any other single form, was seen in its place. Could they fail to see in this a closer, a more intimate revelation, a more exalted symbolism of the Divine Nature and Presence?'

Calendars

The existence of the 6585-day cycle allowing the prediction of eclipses may, obviously, be used to work backwards in time; in other words, to 'hindcast' past eclipses from the exact dates of current eclipses. This technique has been of great value to historians in fixing the dates of certain events, which, from contemporary accounts, are known to have coincided with one or other eclipse. Eclipses calculated in this way provide a way of cross-checking dates and establishing the chronology of events and that of people who witnessed them. The ancient Greek historian Apollodorus of Athens tried to use this method to fix the dates of birth and death of Thales of Miletus, and the astronomer Apianus did the same in the 16th century.

But such reconstructions are not as easy as one might think, not even in this age of computers running high-performance astronomical programs that are capable of calculating, in just a few seconds, the appearance of the sky at any given place and at any given date. The reason concerns the complexity of calendars and also of the rotation of the Earth.

In the first place, it is essential to know the relationship between the dates used by historians and those employed by astronomers. To the latter, the principal object of the chronology is to convert the dates expressed in any particular scheme into the dates in our calendar.

In the Christian West, the historical chronology was established, de post facto, by the early Church Fathers, at a period when zero had no status in mathematics. (There was no way of writing zero in Roman numbering.) The historians did not, therefore, use Year Zero in their dating, so that one jumps directly from the year 1 BC to the year 1 AD, the two years being separated by an 'instant' (and not a year) numbered zero. That is why the 21st century and the third millennium begin on

1 January 2001. This is not very convenient for arithmetical calculations, notably for retaining the property that leap years are divisible by four. In 1740, the astronomer Cassini therefore introduced a numbering scheme based on negative integers to date ancient events. He defined the year 0 as the year 1 BC, the year -99 as the year 100 BC, and so on.

The situation is complicated because several calendars were used in succession. The 15 October 1582 saw the introduction of the Gregorian calendar. That date corresponds with 5 October in the old calendar, known as the 'Julian' calendar, because it was established during the reign of the Roman emperor Julius Caesar.

Ever since the 13th century, the theologians had agreed with the astronomers in recommending reform of the Julian calendar. The Julian year assumed a length for the solar year that was 11 minutes 14 seconds too long. This amounted to a whole day over a period of 128 years, so that the difference between the calendar and astronomical time continued to increase, reaching ten days by the 16th century. In 1582, a bull issued by Pope Gregory introduced the new calendar. The ten days 'too many' fell on the 4th October, so the date went straight to the 15th.

The calendars employed by other peoples, whether the Hebrews, Assyrians, or Chinese, lead us into a veritable jungle in which only specialists in chronology are able to keep track. In general, the Chinese, Assyrian, Alexandrine, Greek, Roman, and other dates are converted into the Julian calendar. They are therefore expressed in years, months and days on this calendar. The major reference points are the Olympiads, the death of Alexander, and the foundation of Rome.

Eclipses and rotation of the Earth

Once a relationship has been established between the various calendars, the reconstruction of past astronomical events is greatly complicated by a purely physical reason: the length of the day is not constant over the course of time! Modern astronomers measure the length of the day by means of atomic clocks, which are used to time the meridian transits of specific astronomical objects. These observations reveal that the Earth's period of rotation varies in a perceptible manner. The Earth is sometimes fast, and sometimes slow, by about 3 milliseconds against the time kept by atomic clocks. Other variations in the day occur over far longer periods. Between 1870 and 1900, for example, the Earth slowed down, causing the length of the day to lengthen by about 7 milliseconds, whereas the tendency reversed in the following 30 years. The cause of these variations is poorly understood; they are probably caused by complex magnetic interactions between the Earth's molten core and the solid mantle.

These variations, which are difficult to model, are superimposed on regular variations with far longer periods. On the one hand, the analysis of the orbits of artificial satellites reveals that the tides raised by the Sun and the Moon in the oceans and, to a lesser extent, in the terrestrial continents, cause a continuous increase in the length of the day by 2.3 milliseconds over a period of a century. On the other hand, the progressive elevation of regions that were covered in ice during the last ice age produces the opposite tendency, and reduces the length of the day by 0.5 milliseconds per century. The net result is a slowing down of the Earth's rotation, which translates into an average increase in the length of the day of 1.8 milli-seconds per century. The cumulative effect over the centuries is considerable. Since the oldest total eclipse of the Sun of which we can be certain, recorded in China in 709 BC, about one million days have elapsed. On average, each of those past days was slightly shorter than a current day, and the cumulative difference amounts to almost 7 hours. At the epoch corresponding to the birth of Jesus, this difference was 3 hours 30 minutes, and it reduces to one hour for the year 1000.

Only the most powerful of professional astronomical programs incorporate these variations in time. For lack of which – as any amateur astronomer can verify on their own computer – an eclipse believed to be total, let us say over Babylon, 3000 years ago, is found to have been shifted over China...

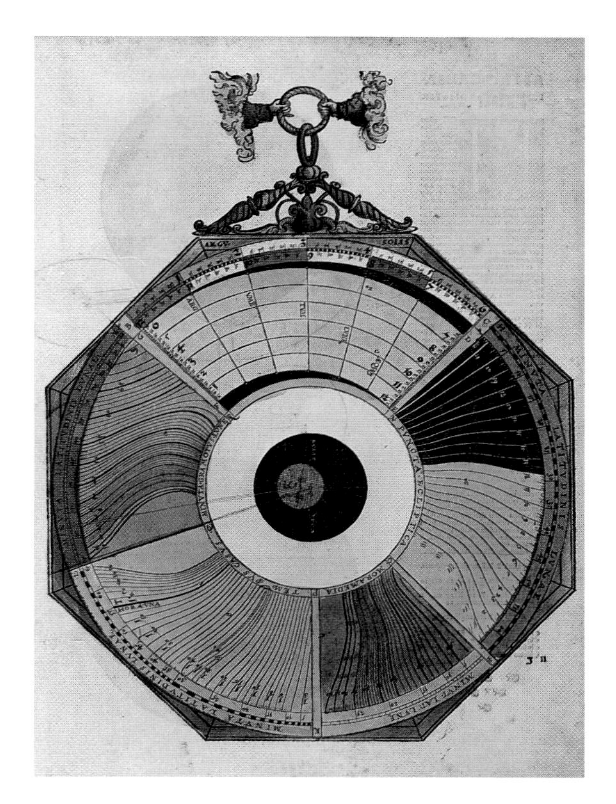

■ This movable disk by Apianus calculates the magnitude, the beginning, and the duration of eclipses over a period of 14 000 years, with remarkable precision.

Conversely, exact observations of past eclipses give the most useful data for determining the long-term changes in the Earth's rotation period. If an eclipse is perfectly dated by historians, it provides astronomers with a valuable indication that enables them to refine their values for the variation in the Earth's rotation. This is particularly the case for the eclipse that occurred in 181 BC, accurately dated in the Chinese annals, which provide the year, month, day, and hour, thanks to the position of the Sun being given relative to specific reference points. Each hour of the day is, in fact, named (*wu*, *shen*, etc.), and the instant that the disk of the Sun is occulted, which marks the exact beginning of the eclipse, is given to the nearest hour, reckoned in *ke*, which are divisions equal to 1/100 of a day, i.e., 14.4 minutes, an interval of time which the

Chinese probably measured with water-clocks. The eclipse was visible at Ch'ang-an (present-day Xi'an), which was then the capital. Modern calculations show that for this eclipse to have been total at Ch'ang-an, the cumulative difference would have to lie between 3.28 and 3.53 hours.

Similarly, the most recent eclipse of 25 June 1275, observed at Hangzhou, imples a difference of 0.36 hours. Most of the Chinese eclipse chronologies date from two distinct periods, 400–600 and 1000–1300, comprising, in total, about 100 observations. The various results obtained from these records, combined with observations made in other parts of the world (Babylonia, Europe, and Arab countries) allow us to draw a curve, which shows that, over the last 2700 years, the average increase in the length of a day has been 1.7 milliseconds per century, and that the actual values fluctuated between 0 and 5 milliseconds per century. The probable causes of this 'variation of the variation' may be ascribed to overall climatic changes, and to core-mantle interactions, which affect the Earth's moment of inertia.

■ THIS ASSYRIAN TABLET DATING FROM THE 2ND CENTURY BC IS THE LIST OF CONTENTS OF THE *GREAT TREATISE OF ASTROLOGY OF THE CITY OF URUK* (IN LOWER MESOPOTAMIA). THE BABYLONIAN CIVILIZATION KNEW HOW TO PREDICT ECLIPSES, THANKS TO THE DISCOVERY OF AN 18-YEAR CYCLE, THE SAROS.

A CANON OF MEMORABLE ECLIPSES

Major past eclipses of the Sun and the Moon are listed in chronological order, and are classed as 'memorable' because they were associated with historical periods or important events. Some of them may even have influenced the course of human history. Their dates are given in the Julian calendar until 1582, and subsequently in the Gregorian calendar. The listing ends in the middle of the 19th century, not because there have been no major eclipses since then, but because from that period onward, most people abandoned their myths and superstitions about such celestial phenomena. Thenceforward, scientists took up the baton, by establishing, partly thanks to eclipses, a new set of legends: that of modern astrophysics, which led to the profound scientific understanding of our Sun, the Solar System, and more generally, of the entire universe itself. This modern legend is described in Chapter 6.

22 OCTOBER 2137 BC, CHINA.
ECLIPSE OF HO AND HI

'Here lie the bodies of Ho and Hi
Whose fate though sad was risible,
Being hanged because they could not
spy Th' eclipse which was invisible.'

An anonymous author thus reports the oldest eclipse in the Chinese annals, going back to the 22nd century BC. According to the legend, the two brothers Ho and Hi, astronomers at the imperial court, were charged with following events in the heavens, defining the seasons, and predicting eclipses. Sadly, they were too fond of the bottle. On the day predicted for the eclipse, dead drunk, they neglected to take the usual precautions and neither summoned the archers nor brought out the drums to frighten the dragon away. The Sun survived, but the Emperor of China had Ho and Hi decapitated for having neglected their professional duties. Since then, some people maintain that no one has ever seen an astronomer drunk on the day of an eclipse, which is, of course, pure fantasy.

The eclipse was probably solar, because the *Shi Jing* manuscript records that 'the Sun and Moon did not meet harmoniously.' But the description was written during a period of about two centuries when there were several total solar eclipses visible in China. The one of 22 October 2137 BC, is a reasonable guess, but the one in the year 2159 BC is equally likely.

3 MAY 1375 BC, MESOPOTAMIA.
THE UGARIT SOLAR ECLIPSE

'On the day of the New Moon, in the month of Hiyar, the Sun was covered in shame and disappeared in the middle of the day, in the presence of Mars.'

Found in the city of Ugarit in Mesopotamia, this inscription mentions a solar eclipse which may have occurred either on 3 May 1375 BC or in 1223 BC.

■ THERE IS A VERY RICH MEDIEVAL ICONOGRAPHY ABOUT ECLIPSES. ALL THE GENERAL ENCYCLOPAEDIAS GIVE A LOT OF SPACE TO EXPLAINING THEM. THIS IS *L'IMAGE DU MONDE*, COMPILED BY GOSSOUIN OF METZ IN 1245.

5 JUNE 1301 BC, CHINA.
THE SHANG DYNASTY ECLIPSE

Oracular texts from Anyang, scratched on animal bones and tortoise shells, date from the Shang Dynasty (c. 1300–1050 BC), and were intended for divination. Dates are rarely indicated, but the solar eclipse of 5 June 1301 BC was mentioned.

15 JUNE 763 BC, ASSYRIA.
THE OLD TESTAMENT ECLIPSE

'And on that day,' says the Lord God, 'I will make the Sun go down at noon, and darken the Earth in broad daylight.'
(Amos 8:9)

■ THIS OTHER VERSION OF *L'IMAGE DU MONDE* PLACES THE MECHANISM OF ECLIPSES BACK IN A COSMOLOGICAL CONTEXT. THE UNIVERSE, CENTRED ON THE EARTH, AND ENCLOSED BY THE SPHERE OF FIXED STARS, IS GOVERNED BY THE UNDERLYING DIVINE UNITY OF THE WORLD.

The corresponding eclipse probably took place on 15 June 763 BC. A cross-reference is provided by an Assyrian historical chronicle, known as the *Eponym Canon*. In Assyria, each year was named after a different ruling official and the year's events were recorded under that name in the Canon. Under the year corresponding to 763 BC, a scribe at Nineveh wrote this simple line: *'Insurrection in the City of Assur. In the month of Sivan, the Sun was eclipsed.'*

Historians have thus been able to use this eclipse to improve the chronology of early biblical times.

735 BC, CHINA.

ECLIPSES OF THE CHU DYNASTY

'The Sun was eclipsed, something that did not augur well. Then the Moon shrank, and now the Sun has also shrunk... It is normal for the Moon to be eclipsed, but for the Sun to be eclipsed is a very bad sign!'

Shi Jing (Book of Odes), *8th century BC*

The Chu Dynasty annals (c. 1050–720 BC) contain few astronomical observations. The text quoted refers, one after the other, to a lunar eclipse and a solar eclipse. The solar eclipse is said to have occurred on the day *hsin mao*, i.e., the first day of

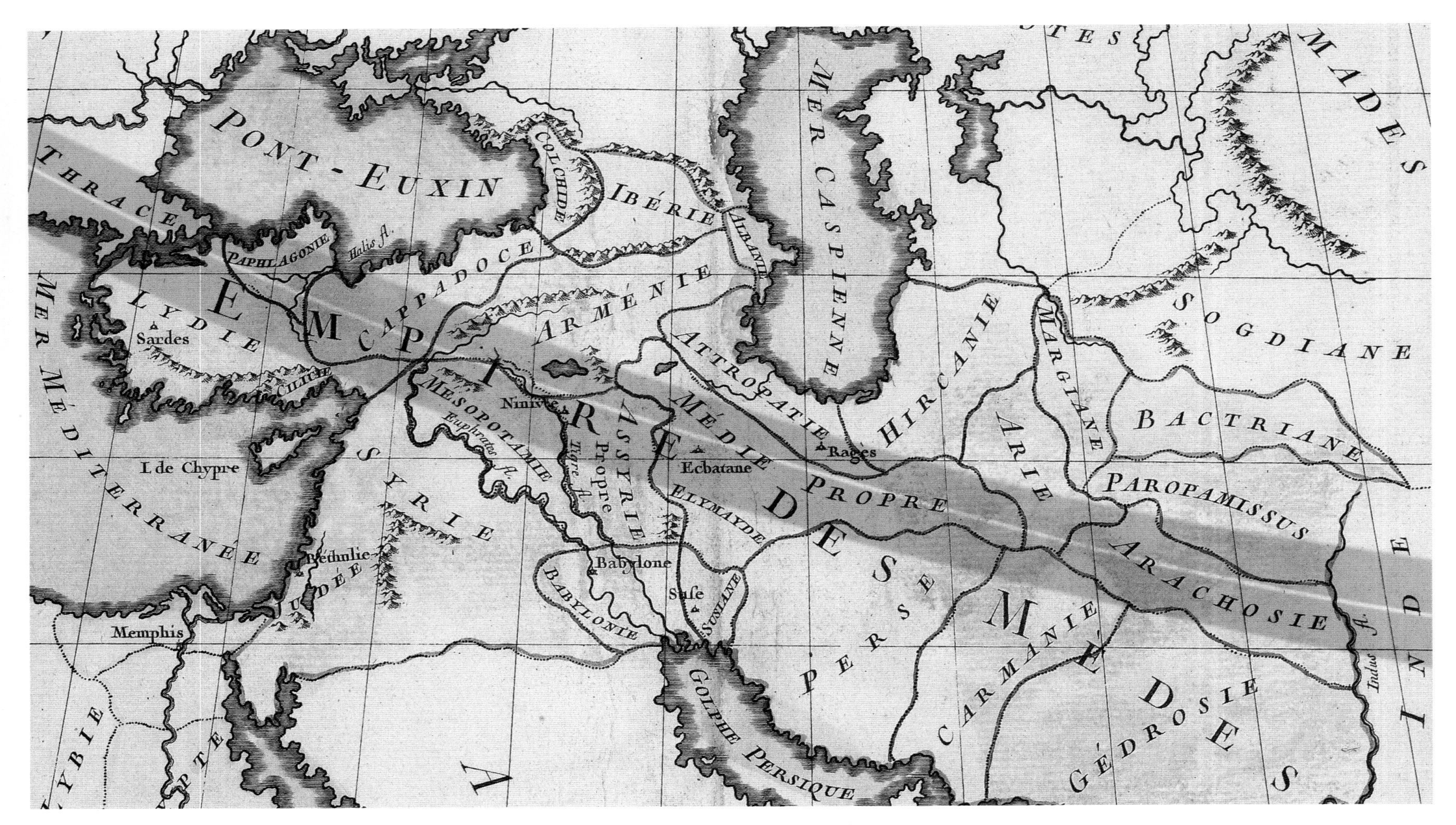

■ In the 6th century BC, the Medes and the Lydians were at war over the frontiers of their kingdoms. A total eclipse of the Sun, which occurred on 28 May 585 BC during a battle, interrupted the conflict, and led to the two kings agreeing to a peace treaty.

the tenth lunar month. Reconstructions suggest that the year in question may have been 735 BC.

17 JULY 709 BC, CHINA.
THE CHUN QIU ECLIPSE.

The *Chun Qiu* (*Spring and Autumn Annals*) covers the period 722 to 481 BC, and contains numerous observations of solar eclipses. In total, thirty-six of them are mentioned, most given to the exact day. Book I, in particular describes an eclipse, stated as being total (*ji*). According to the historian Richard Stephenson, this is the first definite reference to a total eclipse of the Sun from any civilization. Converted to the Julian calendar, the date mentioned agrees exactly with calculations that show a solar eclipse that occurred on 17 July 709 BC. This same eclipse is also referred to in the *Han Shu* (the history of the first Han dynasty), a much later document dating from the 1st century AD.

7TH CENTURY BC, BABYLON.
THE SCRIBES' ECLIPSE

Like the Chinese, the Babylonian civilization carefully recorded eclipse phenomena. The tablets contain numerous cuneiform inscriptions by the scribes, which they signed with their own names. The scribe Bel-suma-iskun, for example, reported a lunar eclipse in 675 BC, and Akkullanu, the solar eclipse of 15 April 657 BC. Rasil the Old indulges in an astrological prediction about the solar eclipse of 27 May 669 BC:

'If the Sun rises like a crescent, carrying a crown like the Moon: the King will capture the enemy's lands; the demon will leave the kingdom, and all will be well.'

6 APRIL 648,BC, GREECE.
ARCHILOCHUS' ECLIPSE

'Nothing there is beyond hope, nothing that can be sworn impossible, nothing wonderful, since Zeus, father of the Olympians, made night from mid-day, hiding the light of the shining Sun, and sore fear came upon men.'

Archilochus, 7th century BC

The spectacle of eclipses impressed the poets of Greek antiquity. In the *Odyssey*, Homer (8th century BC) refers to a solar eclipse that occurred among the various tribulations that beset his hero, Ulysses. The fragment from Archilochus describes a total eclipse which may have taken place on 6 April 648 BC, visible from all of the islands in the Aegean Sea. Archilochus is thus the first Greek poet whose acme (height of activity) may be dated with certainty: between 680 and 640 BC.

28 MAY 585 BC, ASIA MINOR.
THE SOLAR ECLIPSE OF THALES

Thales (c. 625–547 BC), the earliest Greek philosopher, calculated the height of the pyramids in Egypt by comparing their shadows with his own. Born in Miletus, a town in Asia Minor, he is the first representative of the Milesian school, among which two other names stand out: Anaximander and Anaximenes. These thinkers tried to understand the world on the basis of physical and rational explanations, and not on mythical or religious ones. They wondered, in particular, about the constitution of the sky, the nature of the Sun and Moon, and the origin of eclipses. According to his commentators, Thales was the first person to maintain that the Moon did not shine with its own light, but reflected that from the Sun. In antiquity, his fame largely rested on the fact that he predicted a total eclipse of the Sun.

According to astronomers' calculations, this eclipse occurred on 28 May 585 BC. Herodotus recounts that it quickly put an end to a war that had lasted six years between the two 'superpowers' in Asia Minor, the Medes and the Lydians. The warring sides were engaged in a bloody battle when the Moon suddenly hid the Sun. The soldiers immediately stopped fighting, and their commanders interpreted the celestial sign as a warning. Alyattes, king of the Lydians, and Cyaxares, king of the Medes, signed a peace treaty defining the frontier between their two states, and cemented the agreement by a double marriage between their own children.

Plato and, following him, Pliny, attributed real understanding of the phenomenon to Thales: *'The original discovery (of the cause of eclipses) was made in Greece by Thales of Miletus, who, in the fourth year of the 48th Olympiad, predicted the eclipse of the Sun that took place during the reign of Alyattes, during the 170th year after the foundation of Rome.'*

In the 20th century, the poet Francis Ponge still believed this, despite making a mistake in the date of the eclipse: *'But how can I fail to be reminded immediately that Thales was perfectly able to predict the eclipse of 610 BC, and how shall I stop myself from wondering if perchance he did not calculate it, and whether the means of carrying out such a calculation might not be found in Egyptian temples?'*

It is, in fact, highly unlikely that Thales, any more than any other philosopher of his time, should have understood the real cause of eclipses. But he possibly had access to the astro-nomical tables drawn from the empirical observations of the Chaldeans, which indicated the periodic return of eclipses at each Saros.

Other ancient commentators ascribe the understanding of the mechanism of eclipses to one of Thales' pupils, Anaxagoras of Clazomenes (500–428 BC). This original thinker believed that the Sun was a disk of glowing metal, larger than the Peloponnese. For this he was accused of sacrilege, and exiled, despite the protection of his friend Pericles. At that time, in fact, the Sun, the Moon, and the Earth, were considered to be untouchable deities. By wanting to think of them as material objects, Anaxagoras was the first thinker in the historical record to be perse-cuted by the religious authorities.

27 AUGUST 413, BC, SICILY.
THE ECLIPSE OF NICIAS

'All was at last ready, and they were on the point of sailing away, when an eclipse of the moon, which was then at the full, took place. Most of the Athenians, deeply impressed by this occurrence, now urged the generals to wait; and Nicias, who was some-what over-addicted to divination and practices of that kind, refused from that moment even to take the question of departure into consideration, until they had waited the thrice nine days prescribed by the soothsayers.'

Thucydides, *History of the Pelopennesian War*

Thucydides (c. 460–395 BC) is considered to be the greatest ancient historian, for having invented historical critique, and for having discovered twenty centuries before Machiavelli, that the real driving force of history was not morality, but the desire for power on the part of individuals and states. In his account of the Pelopennesian War, Thucydides reports two eclipses that took place during the fight that set Athens against Sparta between 431 and 404 BC. Athens, democratic, inventive, and maritime, was just the opposite of Sparta, which was bellicose, rigid, and land-based. The first eclipse, an annular solar one, occurred during the summer of 431 BC. The other, lunar, eclipse took place on precisely 27 August 413 BC, during the siege of Syracuse. That year, during the second Athenian expedition against Sicily, the fleet commanded by Nicias was trapped for several days in the great roadstead at Syracuse, the exit from which was controlled by the Syracusians. The generals decided to break out with a surprise attack. During the chosen night, the 27th August 413 BC, an eclipse of the Moon occurred. Immediately, Nicias, seeing this eclipse as a sign from the gods, suspended operations, to the great satisfaction of the terrified soldiers. Nicias decided to wait for the next Full Moon. This delay allowed the Syracusians the time to understand the tactics and to reinforce their blockade. When the Athenian fleet finally tried to break out a month later, it was driven back into the bay, and its troops were massacred; 29 000 soldiers perished, and 200 ships were destroyed. This defeat caused the fall of Athens, the cradle of western civilization, which finally collapsed in 406 BC.

This is how Nicias' lunar eclipse had a decisive effect on history.

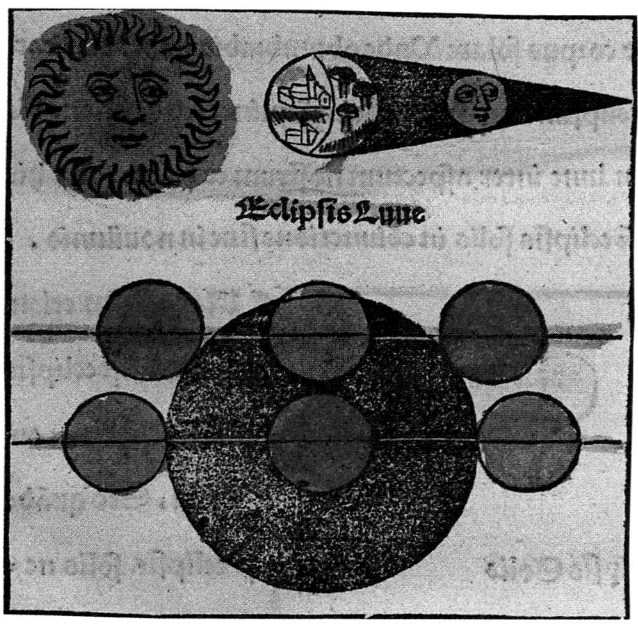

■ This astronomical book by Sacrobosco, Treatise on the Sphere (12th century), was a standard text throughout the Middle Ages and the Renaissance. It gives the explanation of eclipses of the Moon, total or partial, depending on the way in which our satellite crosses the Earth's shadow cone.

15 AUGUST 310 BC, SICILY.
AGATHOCLES' SOLAR ECLIPSE

In his *Bibliotheca historica* (*Historical Library*), Diodorus Siculus (1st century BC) describes how Agathocles, the Syracusian dictator, was able to escape the Carthaginian blockade with his fleet of 60 ships. A total eclipse of the Sun frightened Agathocles' sailors so much that they fled from the famous roadstead during the darkness, and continued onwards for six days and nights. When they finally stopped, they realised that they had scattered the enemy fleet! The event may be dated precisely as 15 August 310 BC.

24 NOVEMBER 29 AD, PALESTINE.
THE CRUCIFIXION ECLIPSE?

'I will show portents in the sky and on earth, blood and fire and columns of smoke; the sun shall be turned into darkness and the moon into blood before the great and terrible day of the Lord comes.'

(*Joel 2:30*)

In *Acts of the Apostles*, Peter also refers to a moon that is the colour of blood and a darkened sky. According to the Evangelists, Jesus was crucified on a Thursday afternoon in the month of Nisan, during the governorship of Pontius Pilate. According to specialists in Roman history, Pilate governed between 26 and 36 (in our current calendar). If we interpret the quotations from the Evangelists as indicating an eclipse that occurred on the day of the Crucifixion, to date the event precisely we need to check eclipses that were visible at Jerusalem during that period. The question is fraught with controversy. According to some, it was an eclipse of the Moon, either that of Thursday, 7 April in the year 30, or Thursday, 3 April in the year 33. Both in fact correspond to the thirteenth day of the month of Nisan, the eve of Passover. According to others, we are dealing with an eclipse of the Sun. One date suggested is 19 March 33 – still during this notable month of Nisan. The problem is that this eclipse was utterly invisible at Jerusalem, as shown by astronomical simulations! The best agreement is that of a solar eclipse visible at Jerusalem on 24 November 29, between 9h 35m and 12h 27m local time. In addition, the Greek historian Phlegon mentions this solar eclipse in his *History of the Olympiads*, and notes that it was accompanied by an earthquake.

24 NOVEMBER 569, NORTH AFRICA.
SOLAR ECLIPSE AT THE BIRTH OF MOHAMMED

The year that the prophet Mohammed was born there was an eclipse of the Sun, where the shadow crossed Sudan and Ethiopia. Farther north, it was only penumbral. In many cultures it was common for eclipses to be associated with the birth or death of the great. Islamic theology does not, however, accept this interpretation.

22 JANUARY 632, ARABIA.
SOLAR ECLIPSE AT THE DEATH OF IBRAHIM

The son of the Prophet, Ibrahim, died tragically on 22 January 632. That day there was an annular eclipse of the Sun, visible in southern Arabia. Some of the devout wanted to see it as a sign from Allah, but Mohammed was specific: *'The Moon and the Sun are signs of God, but they do not go into eclipse for the death or birth of a man.'*

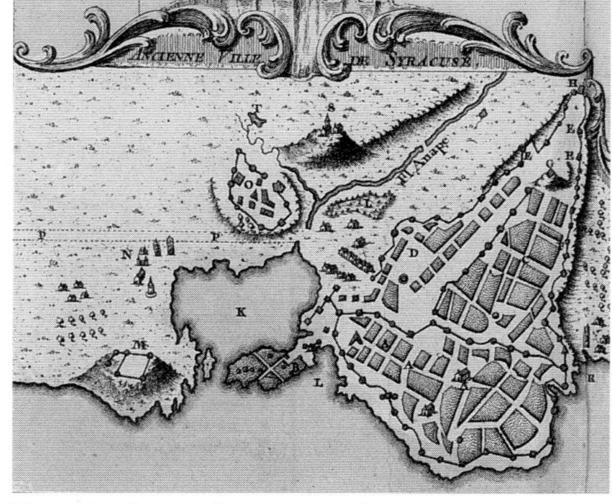

■ THE BAY OF SYRACUSE, IN SICILY, HAS AN EXTREMELY NARROW ENTRANCE. IN 413 BC, THE ATHENIAN FLEET, COMMANDED BY NICIAS, TERRIFIED BY AN ECLIPSE OF THE MOON, REMAINED TRAPPED IN THE PORT. IN 310, ON THE OTHER HAND, AGATHOCLES' FLEET WAS ABLE TO ESCAPE, THANKS TO AN ECLIPSE OF THE SUN.

5 MAY 840, BAVARIA.
SOLAR ECLIPSE OF THE EMPEROR LOUIS

Louis of Bavaria, the son of Charlemagne, was head of a vast empire when, on 5 May 840, he witnessed five minutes of totality during a solar eclipse. He was so petrified that he died shortly afterwards. His three sons then began to dispute the succession. Their quarrel was settled three years later with the Treaty of Verdun, which divided Europe into three large areas, which nowadays correspond to France, Germany, and Italy. The event is reported in the *Chronicles* of Andreas Bergomatis.

6TH TO 12TH CENTURIES, ENGLAND.
ECLIPSES OF THE LATE MIDDLE AGES

The *Anglo Saxon Chronicles* contain numerous references to eclipses visible in England throughout the late Middle Ages: partial solar eclipses in 538 and 540; a total on 1 May 664; an annular visible in the north of England on 14 August 733; another on 16 July 809. Lunar eclipses are also mentioned, often accompanied by the characteristic description of their colour of blood: that of 734, when the Moon 'was as though drenched with blood'; of 23 November 755, when according to Simon of Durham, the eclipsed Moon occulted Jupiter. Other lunar eclipses are reported for the years 800, 806, 829 ('the Moon darkened on Christmas eve'), 1078, 1110, 1117, 1121, and so on.

2 AUGUST 1133, ENGLAND AND GERMANY.
KING HENRY'S ECLIPSE

Visible in England and Germany, this total solar eclipse prompted many descriptions in the chronicles of both countries. For the English, the eclipse took place the day after the departure of King Henry I, and was thus interpreted as a omen of his death. In fact, the king died shortly afterwards in Normandy, which merely subsequently confirmed the superstition. As for the Germans, they associated the darkening of the Sun to the sack of the city of Augsburg and the massacre of its inhabitants by Duke Frederick.

14 MAY 1230.
GREAT SOLAR ECLIPSE OVER EUROPE

'It became so dark that the labourers, who had commenced their morning's work, were obliged to leave it, and returned again to their beds to sleep; but in about an hour's time, to the astonishment of many, the Sun regained its usual brightness.'

This is how the historian Roger of Wendover described this solar eclipse, which occurred early in the morning for western Europe. An understanding of eclipses was not common

■ According to the evangelists, the Sun darkened during the Crucifixion. Later, the event was associated with an eclipse that was visible at Jerusalem. The scene was reproduced many times throughout the history of painting. This is *The Erection of the Cross* by Cornelis de Vos (17th century).

knowledge, reading of encyclopaedias being reserved for priests and the literate.

3 JUNE 1239, SOUTHERN EUROPE.
ARREZO'S SOLAR ECLIPSE

Because it cast its shadow over the whole of southern Europe, this solar eclipse made a deep impression and prompted innumerable accounts, which are found in the archives of the towns of Coimbra (Portugal), Toledo (Spain), Montpellier (France), Arezzo, Florence and Sienna (Italy), and Split (Croatia). The account left by the monk Ristoro d'Arezzo stands out from normal descriptions by its amazing precision: *'While I was in the city of Arezzo, where I was born, and in which I am writing this book, in our monastery, a building which is situated towards the end of the fifth latitude zone, whose latitude from the equator is 42 and a quarter degrees and whose westerly longitude is 32 and a third, one Friday, at the 6th hour of the day, when the Sun was 20 deg in Gemini and the weather was calm and clear, the sky began to turn yellow and I saw the whole body of the Sun covered step by step and it became night. I saw Mercury close to the Sun, and all the animals and birds were terrified; and the wild beasts could easily be caught. There were some people who caught birds and animals, because they were bewildered. I saw the Sun entirely covered for the space of time in which a man could walk fully 250 paces. The air and the ground began to become cold.'*

22 MAY 1453, BYZANTINE EMPIRE.
ECLIPSE OF THE MOON
FALL OF CONSTANTINOPLE

The city of Constantinople, ancient Byzantium, remained the capital of the eastern Mediterranean for a thousand years. By the 15th century, the Ottoman Empire, which was in a state of major expansion, decided to conquer it. The Turks laid siege to it in 1402 and 1422, without success, the city being surrounded by impregnable ramparts. In 1453, the troops of Sultan Mohammed II returned to the foot of the walls. In addition to their 250 000 men, the Turks had a weighty argument: a new cannon 8 metres long, capable of firing 600-kg cannonballs. Despite everything, the city's defenders, scarcely 7000 in number, repelled three assaults and were able to repair their damaged walls each night. They were completely confident in an old prophesy according to which Constantinople would never fall while the Moon was in its crescent phase. On 22 May, the Full Moon was lost in the shadow of an eclipse. According to modern astronomical reconstructions, the Moon entered the penumbra at 16h 36m (local time), plunged into the umbra at 17h 45m, and did not reappear bright until 21h 50m. The morale of the besieged collapsed. Six days later, Mohammed tried a new assault and succeeded, putting the defenders to rout. The sack of Constantinople that followed was a major shock to western civilization.

■ IN 1453, THE CAPTURE OF CONSTANTINOPLE BY SULTAN MOHAMMED II WAS AIDED BY AN ECLIPSE OF THE MOON. THE DEFENDERS WERE SO FRIGHTENED THAT THEY CEASED TO FIGHT.

29 FEBRUARY 1504, ANTILLES.
CHRISTOPHER COLUMBUS' ECLIPSE

During his fifth voyage to the Americas, in 1503, Christopher Columbus was stranded on the island of Jamaica. His damaged flotilla was being repaired in the small Santa Maria bay, and supplies began to run short. In the beginning, Columbus and his crew had been able to obtain food from the natives, in exchange for trinkets and rubbishy goods. As the months passed, the sailors became more aggressive. Some of them mutinied and killed some of the natives to get hold of their food. The Indians stopped supplying the Westerners. With famine threatening, the Spanish admiral came up with an ingenious plan. He had taken with him the *Calendarium* by the astronomer Regiomontanus, published in Nuremberg in 1474. This contained predictions of lunar eclipses for several years. In particular, it predicted a total eclipse of the Moon for 29 February 1504. For the Greater Antilles, the Moon would enter the penumbra at 16h 53m (local time), the eclipse would be total between 19h 18m and 20h 06m, and the Moon would finally leave the penumbra at 22h 29m. In total, six hours of disturbance in the heavenly order gave Columbus the time to arrange a proper drama. That evening, he asked to see the cacique at a time coinciding with the beginning of the eclipse. He announced that the Christians' God did not like the way in which the natives were treating them, and that he had decided to obliterate the Moon as a sign of his disapproval. Hardly had he announced the disappearance of the Moon than the Earth's shadow started to hide the Full disk. Terrified, the natives begged Columbus to restore the light. The admiral replied that he needed to retire to consult his God. He shut himself in his cabin for an hour and fifty minutes; his God happened to be an hourglass which

■ This illustration from Camille Flammarion's *Astronomie Populaire* shows how Christopher Columbus used an eclipse of the Moon to assert his power over the Indians in Jamaica. It was 29 February 1504.

allowed him to time the eclipse. Just before the end of totality, he reappeared and announced that God had given his pardon, and would allow the Moon to retake its place in the sky, provided the Christians were given provisions. The Moon reappeared immediately. Cowed, the natives met the needs of Columbus and his crew, until they were able to complete their repairs and return to Europe.

1500–1530, GERMANY.

ECLIPSES OF *ASTRONOMICON CEASAREUM*

After having taught astronomy to the emperor Charles V, in 1540, Petrus Apianus dedicated to him and to his brother Ferdinand, his masterwork, consequently called *Astronomicon Ceasareum* (*Astronomy of the Ceasars*). The Emperor financed the printing, rewarded Apianus with a large sum of money and appointed him court mathematician. The *Astronomicon Ceasareum* is noted for being a magnificent work both of astronomy and of printing. Hand-coloured, the book makes extremely clever use of movable disks called 'volvelles', which enable the reader to calculate the position and the movement of heavenly bodies with great accuracy. In particular, it contains five volvelle plates on eclipses. Charles was born on 24 February 1500, and was elected Emperor in 1519. Ferdinand, born on 10 March 1503, became King of Bohemia in 1526, King of Rome in 1531 and, after the abdication of his brother Charles, Emperor in 1556. These dates play an important part in the astrological explanations in the book, because the lives of the two monarchs are repeatedly used as examples in the use of the movable disks.

1487	1488	1489
Eclipsis Solis	Eclipsis Solis	Eclipsis Lune
20 2 6	8 17 30	7 17 41
Julii	Julii	Decembris
Dimidia duratio	Dimidia duratio	Dimidia duratio
0 51	0 41	1 45
Puncta septem	Puncta quattuor	

1490	1490	1491
Eclipsis Lune	Eclipsis Lune	Eclipsis Solis
2 10 6	26 18 25	8 3 18
Junij	Novembris	Maij
Dimidia duratio	Dimidia duratio	Dimidia duratio
1 55	1 47	1 5
		Puncta novem

■ At the end of the 15th century, the German astronomer Regiomontanus calculated a precise calendar of eclipses, which Columbus skilfully used on his voyage to the West Indies.

In addition, a paragraph details the use of smoked glass for observing the Sun whilst protecting the eyes. Sailors already used it, but Apianus mentions it for the first time in astronomical literature.

21 AUGUST 1560, PARIS.

THE ECLIPSE 'NEXT FORTNIGHT'

Although eclipses were perfectly predictable, in times when there were troubles, famines, and war, they were seen as unfavourable portents and a source of fear.

■ THE POLISH ASTRONOMER JOHANNES HEVELIUS WAS A SKILLED OBSERVER OF THE SKY. IN HIS CORRESPONDENCE IN 1676, HE DREW HIMSELF WATCHING A PARTIAL ECLIPSE OF THE SUN.

In France, during the wars of religion at the end of the 16th century, announcement of a solar eclipse in 1560 caused general panic. For some, it presaged a second deluge, and for others that the whole world would be in flames. On the express orders of some doctors, hordes of fearful people shut themselves into cellars. Others jostled to get to the confessional. As the decisive moment approached, one country priest, who showed little regard for astronomical accuracy, and who was no longer able to give absolution to all the parishioners who were convinced that their last hour was come, declared that 'there was no need to hurry, because as there were so many penitents, the eclipse had been put off for a fortnight'!

Many more sanguine spirits made fun of these superstitions. Pierre de L'Estoile, for example, on the occasion of the solar eclipse that occurred in October 1605, wrote ironically: 'Several strange and varied maladies were around in Paris at that time, and with the eclipse that occurred on the twelfth of this month, many people were eclipsed who have not been seen since.'

3 MAY 1715.

EDMOND HALLEY'S ECLIPSE

'A few seconds before the sun was all hid, there discovered itself round the moon a luminous ring about a digit, or perhaps a tenth part of the moon's diameter, in breadth. It was of a pale whiteness, or rather pearl-colour, seeming to me a little tinged with the colours of the iris, and to be concentric with the moon.'

This was how the British astronomer Edmond Halley (1656–1742) described the solar corona during the total eclipse of 1715, believing, however, that he had seen the Moon's atmosphere for the first time. The ecliptic digit was a unit of arc equal to the twelfth part of the hidden body. For the Sun, whose apparent diameter is 30 minutes, a digit therefore represented 2.5 minutes of arc.

Halley became famous for having discovered the periodicity of certain comets and predicted the return, after 76 years, of the one that he had observed in 1682. Basing his calculations on the law of universal attraction established by Newton, for the first time he was able to give a rational explanation for these wandering bodies, whose unexpected appearance terrified the populace.

■ THE MANUSCRIPT OBSERVATIONAL NOTEBOOKS OF THE DANISH ASTRONOMER TYCHO BRAHE INCLUDE A PRECISE DESCRIPTION OF A PARTIAL ECLIPSE OF THE SUN, VISIBLE AT PRAGUE IN 1595.

22 MAY 1724, PARIS.

TOTAL SOLAR ECLIPSE OF LOUIS XV

The last total eclipse of the Sun visible at Paris occurred on 22 May 1724. The young King Louis XV, aged fourteen, saw it. The path of its shadow, very similar to that of the eclipse of 11 August 1999, but shifted southwards, crossed England, France, and Germany. It was carefully calculated and mapped, and painters depicted scenes of the crowds of spectators.

27 OCTOBER 1780.

THE NEW WORLD ECLIPSE

During the War of American Independence, the first expedition to

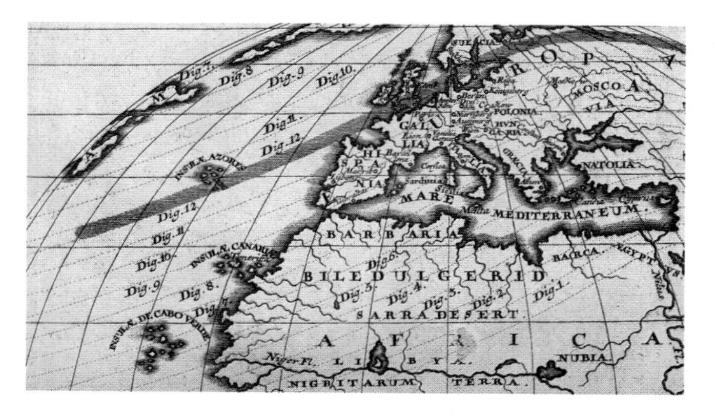

■ THE PATHS OF TOTALITY FOR THE SOLAR ECLIPSES OF 3 MAY 1715 AND 22 MAY 1724, WHICH CROSSED EUROPE, WERE CALCULATED VERY ACCURATELY BY CONTEMPORARY ASTRONOMERS.

observe an eclipse was organised by Harvard College. A special protocol guaranteeing immunity was negotiated with the British, so that the scientists could work without hindrance. The least that can be said is that the first American astronomers were not very expert. When, after a lot of effort, the small party arrived at the observing site, they saw nothing of the eclipse. The site that they had calculated was outside the zone of totality!

12 FEBRUARY 1831, NORTH AMERICA.
NAT TURNER'S ECLIPSE

The Slaves' Revolt in North America had its charismatic figure in the person of Nat Turner. In 1828, this slave-turned-preacher had a vision: he would lead his people to liberty, but he should await a sign from the sky. An annular eclipse of the Sun then occurred on 12 February 1831. Its path of totality swept from Louisiana to Virginia. Turner saw it as a 'black angel' occulting a 'white angel', an intimation of Black overcoming White. The time for revolt had arrived. Several months later, after having murdered his original masters, Turner and his band of insurgents headed for the small town of Jerusalem. Their march was promptly interrupted by militiamen. Most of the slaves were captured and executed. Turner remained in hiding for seventy days, before being taken and hanged.

8 JULY 1842, SOUTHERN EUROPE.
THE ASTRONOMERS' ECLIPSE

Many astronomers moved their bases, and carried out the first modern observations of the eclipsed Sun (see Chapter 6). At Milan, at the instant when the Sun became completely dark, the crowd broke into spontaneous cries of 'Long live the astronomers!'

28 JULY 1851, NORTHERN EUROPE.
ADAMS' ECLIPSE

It was during the course of this solar eclipse that the first photograph was taken of the solar corona, that is, of the Sun's outer atmosphere (see Chapter 6). The best observations were made in Scandinavia. An Englishman, Edwin Dunkin, wrote: *'The prominences were clearly visible, especially a large hooked protuberance. This remarkable stream of hydrogen gas, rendered incandescent while passing through the heated photosphere of the Sun, attracted the attention of nearly all the observers at the different stations.'*

The best account comes from a great astronomer, John Couch Adams. In 1845, he had calculated, at the same time as the Frenchman Le Verrier, the position of an unknown planet lying beyond Uranus, a planet that had actually been discovered through the telescope in September 1846 and named Neptune. At that time, a lot of renowned astronomers had never observed a total eclipse, because these rarely occurred at any given place, and transport facilities were far, far worse than they are today. In his article that appeared in the *Memoirs* of the Royal Astronomical Society, Adams' lyrical style conveys the extraordinary emotion of an astronomer, even an experienced one, who realises that he is a complete novice when he sees this spectacle for the first time:

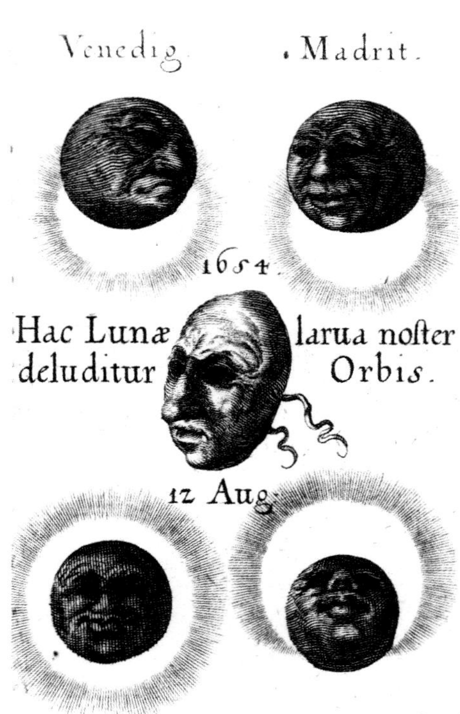

■ J. BALDE DREW A PICTURESQUE REPRESENTATION OF THE SOLAR ECLIPSE OF 12 AUGUST 1654, WHICH WAS OBSERVED IN EUROPE. PARTIAL AT VENICE, MADRID, AND STOCKHOLM, IT WAS ANNULAR AT COPENHAGEN.

'The approach of the total eclipse of July 28, 1851, produced in me a strong desire to witness so rare and striking a phenomenon. Not that I had much hope of being able to add anything of scientific importance to the accounts of the many experienced astronomer who were preparing to observe it; for I was not unaware of the difficulty which one not much accustomed to astronomical observation would have in preserving the requisite coolness and command of the attention amid circumstances so novel, where the points of interest are so numerous, and the time allowed for observation is so short.'

Adams then describes the magical appearance of the corona:

'The appearance of the corona, shining with a cold unearthly light, made an impression on my mind which can never be effaced, and an involuntary feeling of loneliness and disquietude came upon me... . A party of haymakers, who had been laughing and chatting merrily at their work during the early part of the eclipse, were now seated on the ground, in a group near the telescope,

■ The solar eclipse of 1715, visible as partial at Paris, provided the chance to observe the event in various ways: directly through telescopes, smoked glass, pinholes, sieves, and various filters, or indirectly by reflection in a bucket of water.

■ This painting of the solar eclipse of 1724, which was total at Paris, shows how the event aroused the curiosity of the masses, while the professional astronomers studied it from the terrace of the Paris Observatory, which had recently been built by Perrault.

watching what was taking place with the greatest interest, and preserving a profound silence... . A crow was the only animal near me; it seemed quite bewildered, croaking and flying backwards and forwards near the ground in an uncertain manner.'

In another piece, Adams compares the solar corona, that immense circle of rays, with the luminous halo that painters depict around the heads of saints.

1863, CHINA, AND 1885, SUDAN.

GENERAL GORDON'S FATAL ECLIPSES

In the middle of the 19th century, the British general Charles Gordon was charged by the western powers to help the Emperor of China and his Ch'in dynasty in his fight against the Taiping revolt. Gordon commanded an army of Chinese mercenaries and gained victory after victory until, on 25 November 1863, a partial lunar eclipse frightened his troops during the siege of Soochow (Suzhou) in Kiangsu (Jiangsu). The superstitious Chinese took the event for an unfavourable omen for the Emperor, who was thought to be the earthly representative of the celestial order. Soochow was not conquered and the Taiping revolt was settled peacefully. This eclipse was thus the cause of General Gordon's first defeat.

Another eclipse, solar this time, was directly responsible for his death. In 1885, he was in charge of the defence of Khartoum, the capital of the Sudan, which was under attack by a charismatic religious leader, the Mahdi. A solar eclipse took place that demoralized Gordon's native troops. The city was taken before British troops could arrive as reinforcements. The British general did not survive the massacre.

29 JULY 1878, UNITED STATES.

THE PIKE'S PEAK ECLIPSE

While professional astronomers climbed to the summit of Pike's Peak in Wyoming to observe this total solar eclipse, the famous inventor Thomas Edison set up his instruments on a chicken farm. When the Sun started to fade, the poultry became demented and attacked him. The inventor took so much time fighting them off, that he had only a few seconds to observe the eclipse. Yet totality had lasted more than three minutes.

■ Parisians during the solar eclipse of 28 July 1851. The eclipse was partial at the capital, but still intrigued the onlookers.

4 JULY 1917, EGYPT.

LAWRENCE OF ARABIA'S LUNAR ECLIPSE

During the First World War, Thomas Edward Lawrence – better known under the name of Lawrence of Arabia – advised the Arabs in their revolt against the Ottoman Empire, an ally of Germany. One of his greatest exploits was the capture of Aqaba, a fortified port on the Sinai Peninsula, with a small troop of 50 Bedouin. In the Seven Pillars of Wisdom, Lawrence reports how a lunar eclipse helped him to overcome the first defensive position, Kethira:

'By my diary there was an eclipse. Duly it came, and the Arabs forced the post without loss, while the superstitious soldiers were firing rifles and clanging copper pots to rescue the threatened satellite.'

Aqaba was taken a few days later. Thanks to this strategic port having fallen to the British, Jerusalem and Damascus were soon recaptured by the Allies. The Turkish soldiers had another reason to be afraid of the eclipse: according to an Islamic tradition, the Day of the Last Judgement is linked to an eclipse in the middle of the month of Ramadan. This was precisely the case on that date.

29 MAY 1919, BRAZIL AND AFRICA.

THE EINSTEIN ECLIPSE

This total eclipse of the Sun was used to confirm, in spectacular fashion, the new Theory of General Relativity proposed by Einstein in 1915. Measurements proved that the paths of rays of light were deflected by a powerful gravitational field (see Chapter 6). This meant that gravitation was not correctly described by Newton's law of universal attraction, but should be interpreted as an indication of an underlying 'curvature' of space-time, caused by massive bodies. Although the public were completely mystified by the new theory, Einstein became the world's most popular scientist almost overnight. ■

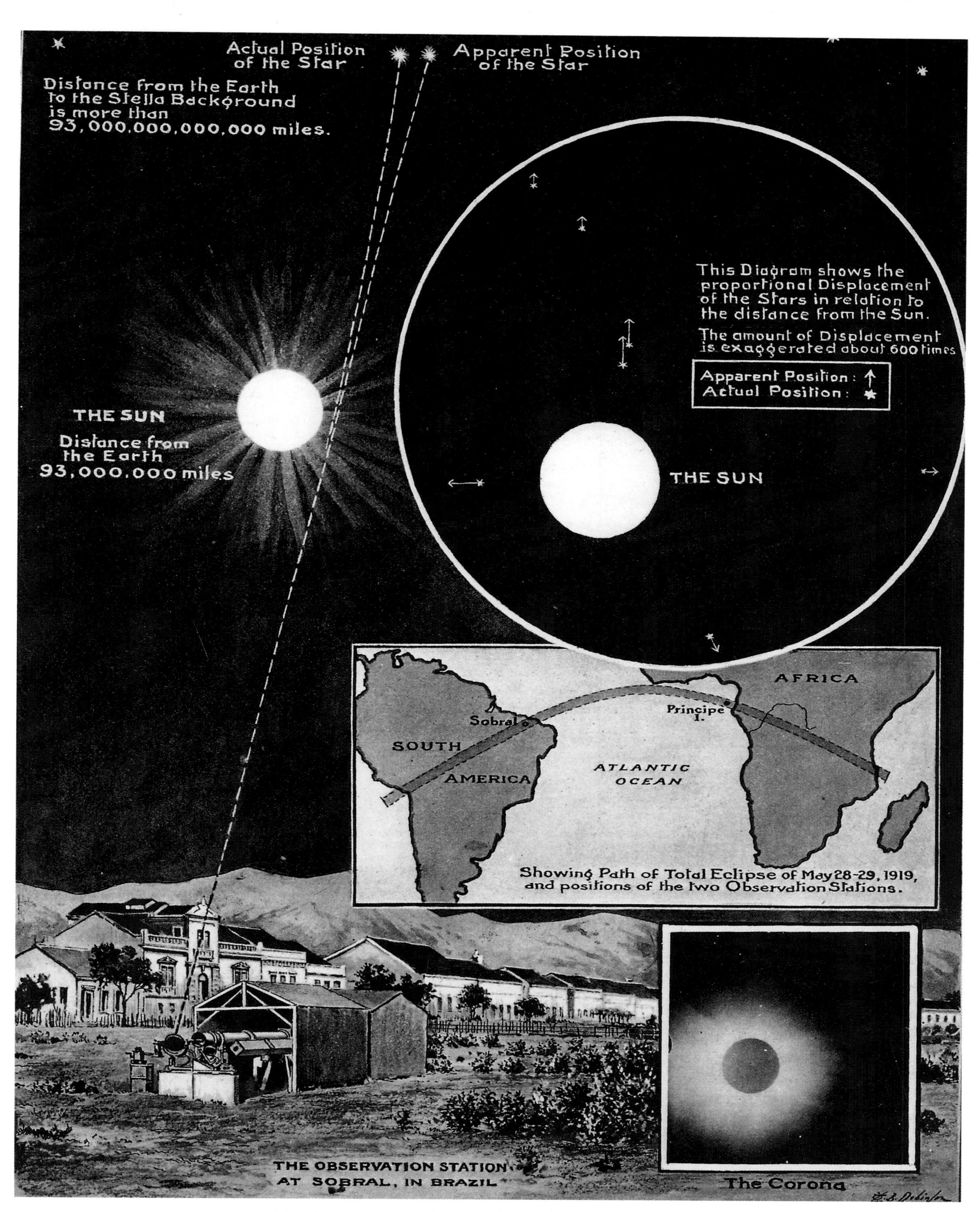

■ This page from the *Illustrated London News* explains the importance of the eclipse of 29 May 1919. Observed in Brazil and Africa, it allowed Einstein's theory of general relativity to be confirmed, thanks to the way the rays of light are bent in passing close to the Sun.

Songs of the eclipses

■ The Sun is in love with the Earth
The Earth is in love with the Sun
That concerns them alone
That is their affair
And when there are eclipses
It is not prudent to watch
Through dirty little pieces of smoked glass
When they are having a row
Those are personal matters
It is better not to become involved
Involved, you risk being changed
Into a cold potato
Or into curling tongs
The Sun loves the Earth
The Earth loves the Sun
That's how it is
Nothing else is our concern

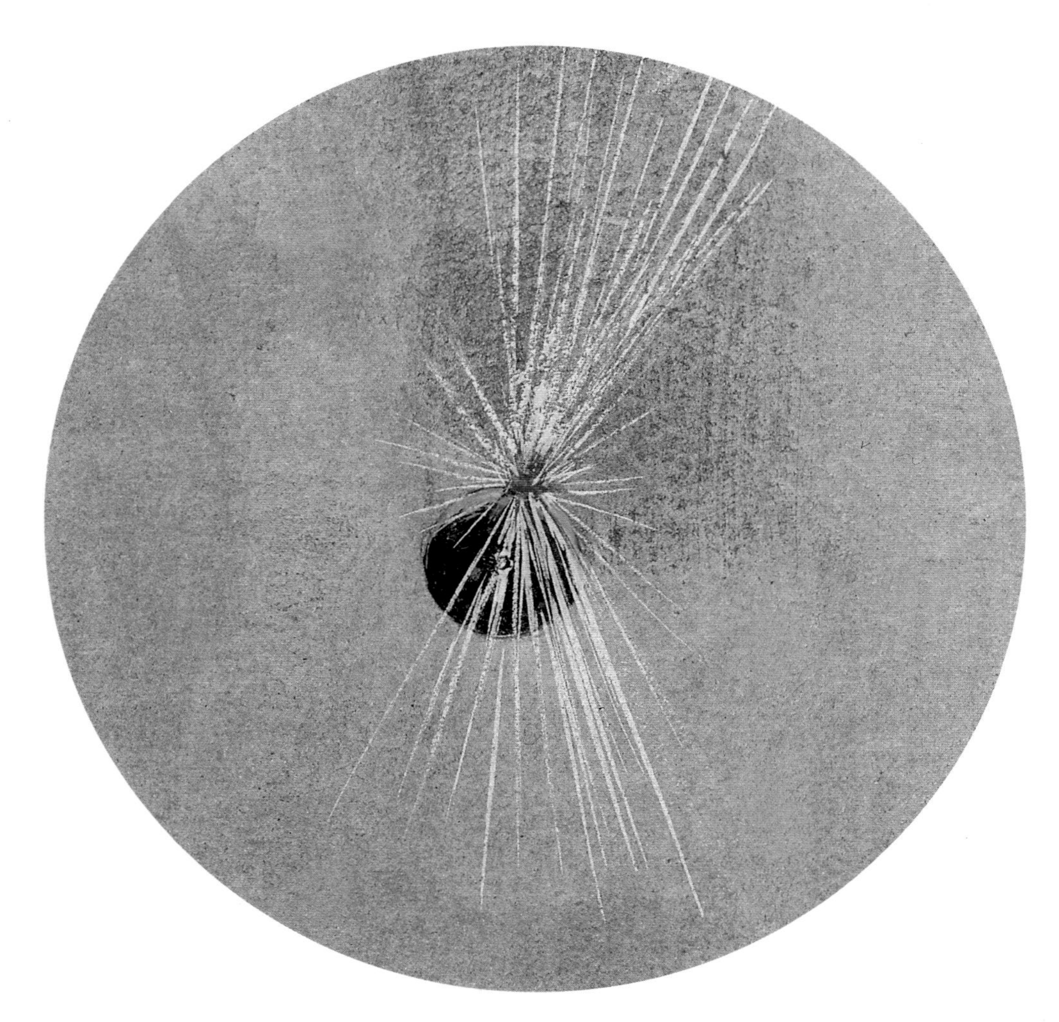

■ THIS ASTONISHING DRAWING BY SAUVE, AN AMATEUR ASTRONOMER WHO OBSERVED THE ECLIPSE OF 1842, SHOWS THE BRILLIANT DIAMOND-RING EFFECT THAT ACCOMPANIES THE REAPPEARANCE OF THE SUN AT THE END OF A TOTAL ECLIPSE.

From Mexico to Babylon, ancient cosmogonies agreed in predicting to their terrified congregations that there would be a long period of darkness, which would seem to reign for ever. Moreover, this period always ended with the rising of a rejuvenated Sun, and the beginning of a new cycle. But the darkness was an inner one. The daily fears with the coming of dusk, which was related to personal experience, was enough to suggest the image of a Sun that might never rise again or which might be extinguished. Because of the link with the secrets of the imagination, eclipses easily became established in literature and poetry. Depending on one's temperament, the disappearance of the Sun or the Moon in darkness is either a pleasant dream or a nightmare. This explains why this dream recurs so frequently, expressed in various ways in the language of poetry, and even in pictorial form in certain mental patients: ominous skies, a darkened Sun, the Moon oozing blood, deathly stars, and landscapes that are a wasteland.

The literature surrounding eclipses is very abundant. Nevertheless it is no more than a subset of a far greater body of work devoted to the Sun and the Moon. Because the Sun and the Moon have a long literary career behind them. 'The damned bitch is poetic!' Mallarmé exclaimed about the Moon, which, in a reaction against romanticism, he had sworn never to invoke in his poetry. Far from giving an exhaustive list here, I shall mention just a few interesting texts on eclipses.

Homer, in the *Odyssey* (written in the 8th century BC), refers to a total eclipse of the Sun, which may have been that of 16 April 1178 BC. In his play, *The Clouds*, written in 423 BC, Aristophanes probably refers to a lunar eclipse that took place in 425 BC and an annular solar eclipse of 424 BC.

The Latin poet Lucretius (1st century BC), who was a disciple of the atomist Epicurus, described his unconventional views of the universe in his master-work *De Natura rerum* (*On the Nature of Things*), about which Francis Ponge wrote, twenty centuries later: 'I have been re-reading Lucretius, and have thought to myself that no one has ever written anything as beautiful.' Lucretius searched, in particular, for the causes of the movement of celestial bodies in accordance with physical laws. On eclipses he wrote:

'And now to what remains!- Since I've resolved
By what arrangements all things come to pass
Through the blue regions of the mighty world,
How we can know what energy and cause
Started the various courses of the Sun
And the Moon's goings, and by what far means
They can succumb, the while with thwarted light,
And veil with shade the unsuspecting lands,
When, as it were, they blink, and then again
With open eye survey all regions wide,
Resplendent with white radiance.'

■ These two artist's impressions, painted by W. Kranz in 1853, illustrate the difference between an annular eclipse of the Sun (*top*), and a total eclipse (*bottom*). Only the latter plunges the Earth into darkness, while a magical light bathes the horizon.

■ This work by Rutherford and Proctor (1873) visualizes what Kepler imagined in his *Somnium* of 1609: Earthshine, and a 'New Earth', as seen from the Moon. The latter is particularly favorable for total solar eclipses, because the apparent disk of the Earth as seen from the Moon is much larger than that of the

Plutarch (c. 50 to c.125), the author of *Parallel Lives* for which he has been called the 'prince of biographers', devoted one of his dialogues, *The Face in the Disk of the Moon*, to the Moon. In a semi-fantastic, semi-serious style, Plutarch has spirits (in Greek *daemon*) crossing from the Earth to the Moon in the shadow-cone that links the two bodies during a lunar eclipse.

■ According to *Genesis*, on the fourth day, God created the heavenly bodies. This engraving is the frontispiece of the poem *La Semaine* (1578) that Du Bartas composed on the creation of the world.

This idea is revived in an astonishing text by Johannes Kepler (1571–1630), *Somnium* (*The Dream or Lunar Astronomy*). Written in 1609, and known at first only in manuscript copies, weighed down by later glosses added by the German astronomer, *The Dream* was not finally printed until 1634, after its author's death, accompanied by a revised Latin translation by Kepler of Plutarch's dialogue. Here again, 'daemons' occupy the Earth's shadow cone and are able to travel, thanks to eclipses, between the Earth and its satellite. But these spirits sometimes take men as their companions:

'We admit to this company nobody who is lethargic, fat or tender. On the contrary, we choose those who spend their time in the constant practice of horsemanship or often sail to the Indies, inured to subsisting on hardtack, garlic, dried fish, and unappetizing victuals. We especially like dried-up old women, experienced from an early age in riding he-goats at night or forked sticks or threadbare cloaks, and in travelling immense expanses of the earth. No men from Germany are acceptable; we do not spurn the firm bodies of Spaniards.

Great as the distance is, the entire trip is consummated in four hours at the most. For we are always very busy, and agree not to start until the moon begins to be eclipsed on its eastern side. Should it regain its full light while we are still in transit, our departure becomes futile.'

The Dream is true space travel, where the explorers are selected and prepared, just like modern-day astronauts, by 'daemons' who closely resemble NASA's own 'sorcerers'.

Jacques Peletier du Mans (1517–1582) is one of the founding poets of the movement known as the 'La Pleiade'. Having participated in the scientific discussions of his time, he maintained that the French language and poetry should be enriched with words borrowed from the vocabulary of science. His poem, *À ceux qui blâment les mathématiques* (*To those who blame mathematicians*), eulogizes mathematicians of the heavens, who are able to predict the occurrence of eclipses and the paths of the planets:

'Was it then nothing to perceive
By courses that were ordinary
The eclipse that would receive
One or other luminary?
To have known, by true mechanics
To calculate those places one by one
And to know for those Erratics
Forwards, backwards, they would run?'

Poet, diplomat and soldier (killed in battle), Guillaume de Salluste, Seigneur Du Bartas (1544–1590), was an imitator and rival of Ronsard, with, however, less metre and less taste. Becoming famous for *La Semaine* (*The Week*), published in 1578, an epic adaptation of *Genesis*, he was flattered by his rival with the words; 'Du Bartas has done more in one Week than I have done in my whole life.' In the Fourth Book (counting the day that the Moon was created in accordance with *Genesis*), Du Bartas becomes didactic, and describes in minute detail the phases of the Moon and the mechanism of eclipses. He explains, in particular, why lunar eclipses are frequent and visible from almost anywhere, while solar eclipses are rare for any given observing site. Du Bartas addresses the Moon; and the Sun is equated with the god Phoebus:

Yet it befalles, even when thy face is Full,
When at the highest thy pale Coursers pull,
When no thicke maske of Clouds can hide away
From living eyes, thy broad, round, glist'ring ray,
Thy light is darkned, and thine eyes are seeld,
Cov'red with shadow of a rustie shield.
For, thy Full face, in his oblique designe
Confronting Phoebus *in th'*Ecliptike *ligne,*
And th'Earth betweene; thou loosest for a space
Thy splendor borrowed of thy Brothers grace:
But, to revenge thee on the Earth, for this
Fore-stalling thee of thy kinde Lovers kisse,
Sometimes thy thicke Orbe thou doo'st inter-blend
Twixt Sol *and us, toward thy later end:*
And then, because his splendor cannot passe
Or pierce the thicknes of thy gloomie Masse,
The Sunne, as subject to Deaths pangs, us sees not,
But seemes all Light-les, though indeed he is-not.
Thearfore, farre diff'ring your Eclipses *are;*
For thine is often, and thy Brothers rare:
Thine doth indeed deface thy beautie bright,
His doth not him, but us bereave of Light:
It is the Earth, that thy defect procures,
It is thy shadow, that the Sunne obscures:
East-ward, thy front beginneth first to lacke;
West-ward, his browes begin their frowning blacke:
Thine, at thy Full, when thy most glorie shines,
His, in thy Waine, when beautie most declines:
Thine's generall, towards Heav'n and Earth together;
His, but to Earth, nor to all Places neither.

William Shakespeare, in *King Lear*, has the Earl of Gloucester say: 'These late eclipses of the sun and moon portend no good to us ... in cities, mutinies; in countries, discord'. The references are to the lunar eclipse of 27 September 1605 and the solar eclipse of 12 October in the same year.

Alongside his scientific works, the Jesuit, Ruggero Giuseppe Boscovich composed a didactic poem, entirely devoted to eclipses. His poetic talents are certainly not on the same level as his scientific understanding. This is how he describes the elliptical shape of the Moon's orbit:

'While eternal revolutions transport Phoebe around us, subjected to general laws she observes the rules common among wandering stars. Her altitudes vary, her unequally curved orbit pulls in its sides, elongates its axis, and resembles the curve engendered by the oblique section of a column.'

■ Antoine Caron, a representative of the French School in the 16th century, painted this scene of ancient astronomers observing a solar eclipse.

In the last canto, Boscovich correctly explains the phenomenon of the reddish Moon: when our satellite is eclipsed by the shadow of the Earth, the reflection and refraction of rays of light in the terrestrial atmosphere give it a reddish hue.

In *Prometheus Unbound*, the English poet Percy Bysshe Shelley (1792–1822) has the Earth and Moon indulge in a burning love duet. The idea of comparing celestial matters with a love story was inspired by the theory of universal attraction developed by Isaac Newton at the end of the 17th century, which explained the motions of celestial bodies subject to the effects of gravity. It was but a small step from the mutual physical attraction between celestial objects to the emotional attraction between human beings, and one that the poets, always in search of metaphors, were happy to embrace. Shelley's text harks back to the Tahitian tradition, that of the celestial lovers who meet to make love when their disks coincide precisely:

'The Earth:
I spin beneath my pyramid of night
Which points into the heavens, dreaming delight,
Murmuring victorious joy in my enchanted sleep;
As a youth lulled in love-dreams faintly sighing,
Under the shadow of his beauty lying,
Which round his rest a watch of light and warmth doth keep.

The Moon:
As in the soft and sweet eclipse,
When soul meets soul on lovers' lips,
High hearts are calm, and brightest eyes are dull;
So when thy shadow falls on me,
Then am I mute and still, by thee
Covered; of thy love, Orb most beautiful,
Full, oh, too full!'

Inspired by the story of Christopher Columbus, who cleverly used an eclipse to gain a vital advantage, Sir Henry Rider Haggard adapted the episode in his famous novel *King Solomon's Mines* (1885). The protagonist interprets an eclipse of the Moon as a sign that his native servant is actually the long-lost king of the Kukuana tribe, which guards a fabulous diamond mine. Rider Haggard's astronomical knowledge left something to be desired, because in the original manuscript, he did not even set the eclipse on a night of Full Moon! This was pointed out to him and he corrected it in the definitive edition. Despite that, no eclipse took place in Africa over the period in which this romantic fiction was set.

The American novelist Mark Twain equally drew inspiration from the Columbus episode in *A Connecticut Yankee at the Court of King Arthur* (1889). The hero, Hank Morgan, is carried back into the past, to the time of mediaeval England, after having had a heavy blow to the head (sic!). Arriving from nowhere on 19 June 528, he is immediately tried as a wizard and condemned to the stake. He knows, however, that on the day set for his execution, there would be a solar eclipse. Pretending to make use of his magic powers, he calls for the event and, during the period of totality, promises to let light return in exchange for his liberty. King Arthur accepts and, naturally, the Sun reappears. Mark Twain gives 21 June 528 as the date of the eclipse. There was, however, no eclipse then, nor on any neighbouring date.

Hergé, the author of the adventures of Tintin, was more scrupulous about the scientific accuracy in his strip cartoons. In *The Temple of the Sun* (1940), Tintin, Captain Haddock and Professor Tournesol are prisoners of a tribe of Incas and are to be sacrificed to the Sun God Pachacamac. The day god will light the pyre by shooting his rays through a magnifying glass. The event occurs at the precise time when an eclipse is due to occur.

Tintin knows about this, having read about it in a newspaper that fell out of Captain Haddock's pocket. It is easy for him to invoke Pachacamac, asking him to mark his disapproval of the execution by veiling his face. The Sun in fact darkens, and the terrified Incas, implore Tintin to ask Pachacamac to return, which he does. Tintin and his friends are released and are able to embark on new adventures. Hergé may have been inspired by the total solar eclipse of 1919 (the 'Einstein eclipse'), whose line of totality crossed Peru and Brazil.

Jacques Prévert warns all solar-eclipse 'voyeurs':

'The Sun is in love with the Earth
The Earth is in love with the Sun
That concerns them alone
That is their affair
And when there are eclipses
It is not prudent to watch
Through dirty little pieces of smoked glass
When they are having a row
Those are personal matters
It is better not to become involved
Involved, you risk being changed
Into a cold potato
Or into curling tongs
The Sun loves the Earth
The Earth loves the Sun
That's how it is
Nothing else is our concern'

The most moving text is the account of a lunar eclipse, written by Chappe d'Auteroche (1728–1769). This French astronomer left on an expedition to Baja California, in Mexico, to observe the transit of Venus across the disk of the Sun on 3 June 1769. The observation was perfectly successful, but meanwhile, a terrible contagious disease decimated the small colony at San José del Cabo. Chappe, wrapped up in his professional zeal, didn't want to know. Once his mission was accomplished, he saw the ravages that his stubbornness had brought about. For two weeks, Chappe did not stop, attending to, and consoling the dying. He, in turn, fell ill. His assistant, the engineer Pauly, begged him to leave. Chappe flatly refused, being overcome with remorse. But he also impatiently awaited the morning of 18 June. That day, a lunar eclipse was expected. With great difficulty, the astronomer hauled himself to the top of the bell tower of the San José Jesuit mission. Being perfectly aware that this would be the last observation of his life, he took particular pains to make it. For the four hours that the event lasted, he wrote down his observations, his hand shaking with fever:

'10h 45m: the eye can see that the Moon has entered the penumbra.

At 11h 08m, the eclipse began. The shadow is so sharp and the Moon so clearly terminated, that I think I may have estimated this contact too late. The craters enter the shadow one after the other: Grimaldus, Galileus, Gassendus, Keplerus, Aristarchus, Tycho. Is it not the lot of astronomers to enter the shadows thus; and shall I, who will never achieve the fame of those I have mentioned, shortly be seized by that great shadow, and blotted out for ever in oblivion?

At 12h 35m we can see half-a-dozen faint stars of the eighth magnitude close to the Moon.

At 12h 54m the edge of Moon darkened slightly more, but it is obvious that the Moon is not yet fully eclipsed, and it appears that the eclipse will not be total.

13h 05m, the illuminated portion increases; Kepler, Copernicus, etc., emerge. The Moon's limb is beginning to brighten.

End of the eclipse at 14h 41m 20s.

At 14h 46m, the limb is almost as clear and well-defined as the rest of the disk, with the difference that it has become a colour that verges on jonquil, when seen towards the edge of a telescope's field of view, whereas the rest of the disk is blue, which proves that there is still some penumbra that can be seen visually on the Moon.'

Once his report had been finished. Chappe asked to be bled by his interpreter. The latter managed, clumsily, to draw a few cups of blood. The blood-letting only exacerbated the disease. Chappe died on 1 August, after having declared that he had fulfilled his contract and that he would die happy. The expedition had two survivors, one being the engineer Pauly. Returning to Paris in 1770, the latter gave Chappe's notes to Cassini, the Director of the Paris Observatory. ■

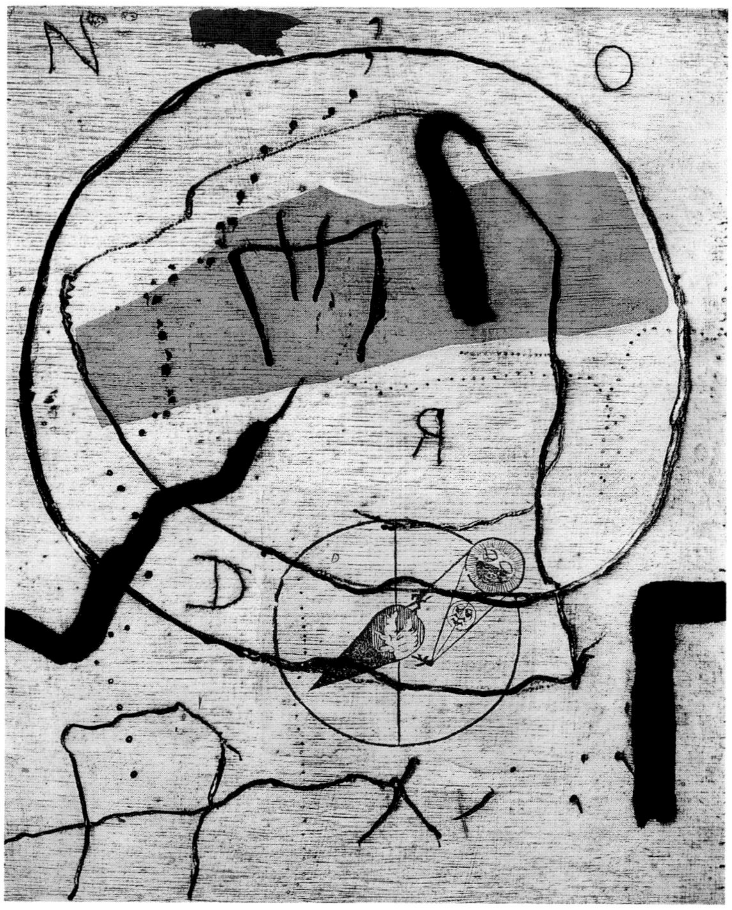

■ *ECLIPSE*, A PAINTING BY RICHARD TEXIER (1991). THE ARTIST, NOTED FOR BEING THE AUTHOR OF *JARDIN DE LUNE*, THE TALE OF AN ASTRONOMICAL GARDEN, HAS INCORPORATED A MEDIEVAL DIAGRAM BY SACROBOSCO INTO HIS PICTURE.

The dance of the Sun and Moon

■ Central eclipse of the Moon. In this photographic montage, the pale disk of the Moon becomes progressively immersed in the cone of shadow cast by the Earth, and then emerges from it about an hour later. At maximum eclipse, our satellite takes on a reddish tinge, caused by refraction of sunlight through the Earth's atmosphere.

■ THE SUN, CAPTURED IN X-RAYS BY THE ASTRONOMICAL SATELLITE TRACE ON 10 OCTOBER 1998. SHOWS FRENETIC ACTIVITY IN THE FORM OF SPOTS AND GIANT ERUPTIONS OF PLASMA.

In *Institutions astronomiques* (*Astronomical institutions*), the first astronomy book written in French, and published in 1557, Jean-Pierre de Mesmes proposed a scientific vocabulary that was drawn from the common language, rather than from Greek or Latin. He suggested that solar eclipses should be called 'obstructions', and lunar eclipses 'fades'. Although these poetic terms were not adopted, they do illustrate the basic difference between the two phenomena rather well.

Although every eclipse is caused by a body (the Moon or the Earth) intercepting light from the Sun, solar and lunar eclipses differ on several points. During a solar eclipse, the Moon masks the Sun either totally or in part, but only for certain points on the surface of the Earth. For some, along a long, narrow track, it is total or annular; for others, it is merely partial, and the hidden portion of the Sun is larger or smaller in size. During a lunar eclipse, by contrast, our satellite ceases completely or partially to be illuminated by the Sun because it crosses the Earth's shadow, and the appearance of the Moon is the same for all the inhabitants of the terrestrial hemisphere for whom the Moon is above the horizon.

It would, therefore, be more accurate to speak of an 'occultation' when the disk of the Moon prevents the light from the Sun from reaching us, and of an 'eclipse' when the Moon's light fades in the Earth's shadow.

A PORTRAIT OF THE ACTORS

However we think of it, eclipses involve three actors with very distinct rôles: the Earth, the Sun, and the Moon. A solar eclipse occurs when the Moon passes in front of the Sun at New Moon; and a lunar eclipse when the Moon passes through the Earth's shadow at Full Moon. In both configurations, the three bodies are aligned.

These bodies have always occupied pride of place in our mythology. Plutarch, for example, combined Plato's cosmology and the celestial religion of the Pythagoreans to construct a theory of personal health, according to which Man consists of a body, a soul, and a power of reason. When the body dies, its soul travels to the Moon, where, still linked with reason, it remains for a certain time in the form of a 'daemon'; after which there is a second death: reason separates from the soul and flies towards the Sun, where it remains for all eternity.

Our physical knowledge of these three bodies has, of course, evolved greatly. In the light of the findings of modern astrophysics, let us paint a quick portrait of the actors in an eclipse.

The Sun is the closest star to us. Its diameter is 1 392 000 km (109 times that of the Earth). Consisting of 99% hydrogen and helium, the Sun has a surface temperature of 5500°C, while at the centre the temperature climbs to 15 million degrees. The latter, together with the enormous pressure that prevails in the

■ The Earth and the Moon are a pair, linked by gravity, separated by a distance, that averages 384\,000 kilometres. Here, the Earth-Moon system appears to the Galileo spaceprobe as two quarters, illuminated by the Sun, some 150 million kilometres away.

■ *The Moon at First Quarter*. The maps of the Moon engraved by Claude Mellan in 1653, at the request of Pierre Gassendi and Nicolas de Peiresc, are now extremely rare. The artist obtained remarkably realistic images thanks to a telescope constructed from optical elements that were provided by Galileo.

core, initiates thermonuclear fusion which converts hydrogen into helium and liberates energy. This energy passes through the gaseous layers of the Sun and propagates out into space; it is the source of the light and heat that created life on Earth, and which are essential to its continuing existence. A large part of solar physics has been acquired thanks to observation of the solar atmosphere during eclipses (*see* Chapter 6). The Sun is a star of average size and mass, belonging to the Milky Way, our own Galaxy. The latter includes about 200 billion stars, some similar to our Sun, and others that are completely different. Some of the stars also possess planetary systems.

The third planet out from the Sun, after Mercury and Venus, the Earth is a sphere, slightly flattened at the poles, with a diameter of 12 750 km at the equator. It orbits the Sun at an average distance of 149 600 000 km, at the rate of one revolution per year, and it rotates on its own axis at the rate of once per day.

Three-quarters of its surface is covered by water. Its oceans and atmosphere give it its characteristic blue colour when seen from space. Its crust mainly consists of silicates: silicon and oxygen together represent 75% of the mass and 95% of the volume of the Earth. Its core mainly consists of molten iron and nickel, with the interior heat arising from the decay of radioactive elements.

The age of the Earth, like that of all the other bodies in the Solar System, is estimated at 4.56 billion years.

The Moon is the Earth's only natural satellite. Its diameter is 3476 km, which is slightly more than a quarter of the Earth's diameter (27%, to be precise). The mass of the Moon is about one eightieth of that of the Earth, which means that on its surface weight is one sixth of that found on our planet.

There is no atmosphere on the Moon. This is why we can see the surface so clearly: shadows are extremely dark and sharply defined. At any given point on the lunar surface there is no dawn or dusk; either blinding light or total darkness reigns, depending on whether the Sun is above or below the horizon. A small amount of sunlight, reflected from the Earth, nevertheless does reach the lunar surface: this is the phenomenon known as *Earthshine*.

Because there is no atmosphere on the Moon, the temperature changes abruptly between day and night. Beneath the Sun, it reaches 130°C. One metre the other side of the terminator – i.e., the line that separates the illuminated part for that in darkness – it falls to 0°C. At lunar midnight, it drops to –180°C. This enormous fluctuation in temperature is possible because the lunar surface absorbs all the heat from the Sun, and because there is no atmosphere to absorb any of it. On Earth, the atmosphere transmits light in the visible region; part of the energy is absorbed, and part is reflected by the oceans and the polar caps. The energy reaching the ground is absorbed and re-emitted as infrared radiation, to which the atmosphere is far less transparent. The energy is thus trapped between the ground and the lower atmosphere and, after sunset, helps to maintain a more pleasant temperature than if there were no atmosphere at all.

When observing the surface of the Moon, we can see that it is scattered with dark grey areas intermingled with areas that are much lighter. Through a telescope, the dark regions appear smooth, whereas the light areas are covered in circular structures, known as craters, and high regions of broken relief, known as highlands. Some craters are minute, with sizes of just a few millimetres; others are gaping holes measuring as much as 240 km across. The edges of these large 'ring plains' are ramparts that may reach as high as 6000 m. A number of the ring plains also have central peaks, and like spiders in the centre of their webs, are surrounded by long radial lines known as rays. The highlands are chains of mountains, some of which are comparable in height to the Himalaya on Earth.

The vast dark, flat areas are called maria (seas). The term was used because observers initially thought that they were vast oceans. But these areas, like all the rest of the Moon, are completely devoid of water.

Galileo Galilei, who made the first observations of lunar relief using a telescope in 1610, and the selenographers of the following two centuries gave poetic names to the lunar 'seas', such as Mare Serenitatis (Sea of Serenity), Mare Tranquillitatis (Sea of Tranquillity), where the first astronauts landed, or Mare Imbrium (Sea of Rains). The craters are named after great scientists and natural philosophers, such as Plato, Ptolemy, Copernicus, and Kepler.

Practically all the structures visible on the lunar surface – craters, maria, and rays – were formed by the

After the discovery of sunspots at the beginning of the 17th century, speculations about the nature of the Sun's surface were rife. Athanasius Kircher notably imagined mountains and lakes of molten lava.

impact of large meteorites. The meteorites, which reach the surface of the Moon at velocities of between 19 and 110 km per second, disintegrate immediately. The tremendous shock causes the ejection of millions of tonnes of rock, leaving behind a circular crater. When the ejecta fall back to the surface, they pile up in a ring around the crater and form the rampart. Some of the rock is fragmented by the explosion into a fine powder, which when deposited on the surface, forms the pale-coloured rays. A rebound from the shock of the explosion may also cause the central peak that is found in the largest craters. The maria were created when giant meteorites, more than 80 km across, crashed into the Moon, early in its history. These impacts did not just create vast craters and high mountains, but their shock-waves cause subterranean rocks to melt. The magma, which then rose to the surface through faults and crevasses, formed the dark plains of solidified lava that we can see today.

Despite all the scientific knowledge that has been acquired about our satellite, astrophysicists stumble over the problem of its origin. Three theories were considered at one time. One maintained that the Moon was once a body orbiting the Sun that approached the Earth, and was captured by it (the capture theory). Another suggested that the Earth and Moon were twin bodies, born side by side from the primordial nebula of dust and gas (the twin planet theory). According to a third hypothesis, the Earth once rotated much faster than today, and it formed a bulge that then became detached to form the satellite (the fission theory). The Russian and American missions that explored the Moon in the 1970s provided information about the material forming our satellite, but did not support any one of these hypotheses. A new theory then saw the light of day: the Moon was formed from debris torn from the Earth by the impact of a body the size of Mars. The collision occurred 4.5 billion years ago, shortly after the formation of the terrestrial core and mantle. The impactor would have vaporized millions of cubic kilometres of rocks from the surface of the Earth and would be disrupted itself, with much of the material being ejected into space. The cooling debris would have gradually accumulated through gravity, to create the Moon as we see it today. The latest gravimetric data obtained by the automated Lunar Prospector probe seem to confirm this model: they have found concentrations of mass (mascons) that underlie vast regions, such as the Mare Imbrium, that arose through the impact of giant meteorites.

Whatever the truth, it is certain that during the course of this primordial period, which lasted 700 million years, the Solar System was subjected to an intense meteoritic and cometary bombardment, with some of the moons having nearly been destroyed, and others broken up only to be subsequently reassembled.

Lunar cycles and their vagaries

When the Moon is visible, it is, after the Sun, the brightest object in the sky. It may be seen in daylight almost as often as at night. But the Moon does not emit any light of its own. It reflects sunlight, and that quite poorly, because its *albedo*, which measures the proportion of reflected light, is only 0.12.

Because the Moon shines by reflected sunlight, only the side facing the Sun is illuminated, whereas the opposite side is dark. This, together with the orbital motion of the Moon, explains why our satellite shows different lighting conditions, known as phases. The explanation of *phases* by the fact that the Moon is solid and illuminated by the Sun goes back to distant antiquity. Thales and Anaxagoras are said to have recognized it, and the Pythagoreans and Cicero described it in detail.

It is through observation of the Moon and its regular phases that have led humanity to adopt the lunar month, or *lunation* – the time the Moon requires to go through all its phases – almost universally as a unit of time. Its precise length is 29.53059 days.

Another extremely important lunar cycle is the *sidereal month*, the time the Moon requires to return to the same place in the sky relative to the fixed stars. This occurs in 27.32166 days. It is easy to understand why this is slightly shorter than the lunation: the cycle of phases involves the relative position of the Sun. After a period of one month, the Earth has moved along its orbit around the latter.

The four main phases are: (1) New Moon, (2) First Quarter, (3) Full Moon, and (4) Last (or Third) Quarter. Because these four phases are observed over 29-and-a-half days, each one lasts about 7 days 9 hours, which explains the approximate division of the lunar month into four weeks. During the two weeks that follow New Moon, that is, the moment at which it is not visible in the sky, the Moon is said to be

■ Johannes Hevelius spent four years mapping the surface of the Moon, baptising 'continents' and 'islands', and completing the most beautiful lunar atlas of all time. This luxurious example of his *Selenographia* was specially coloured for king Louis XIV.

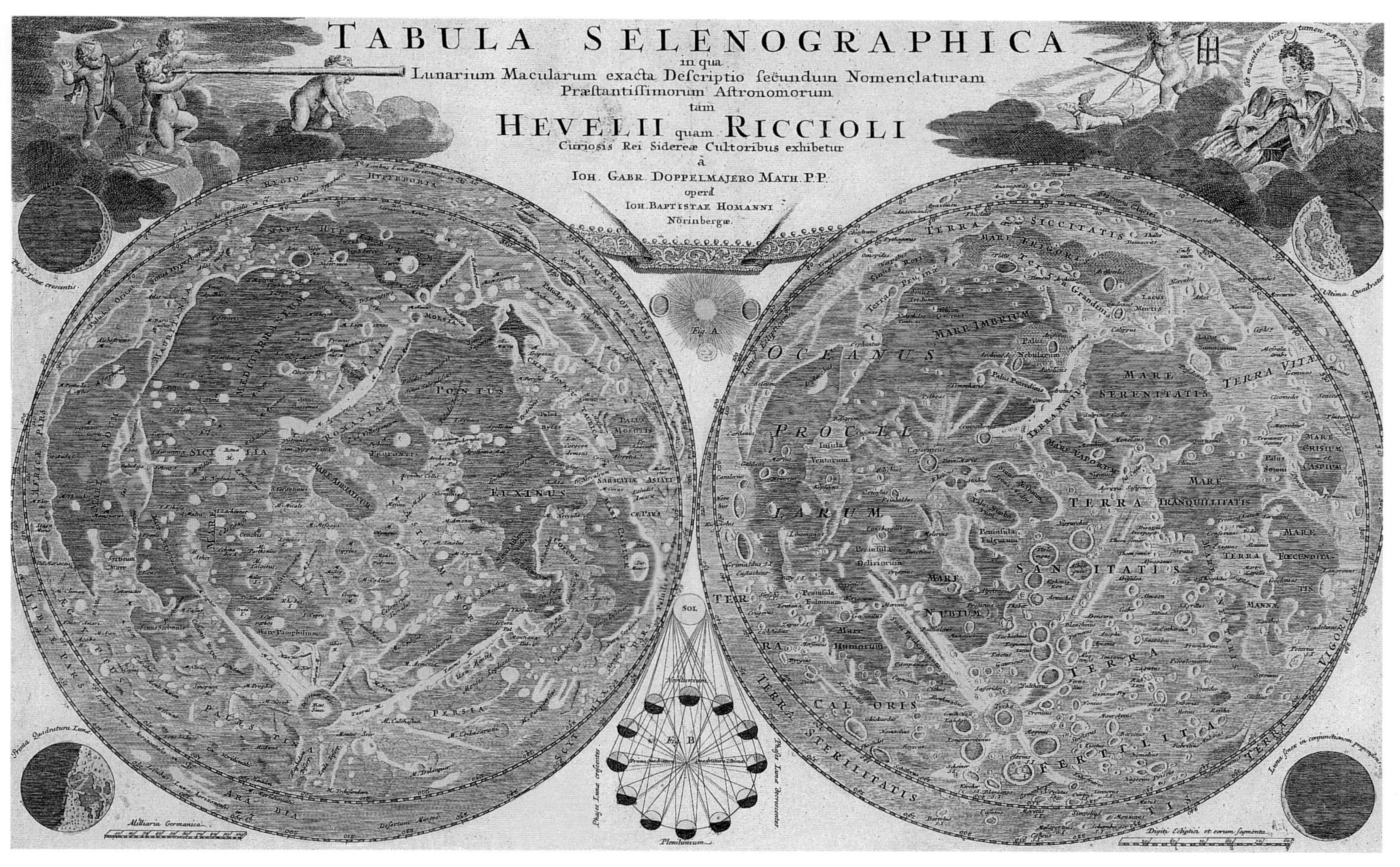

■ IN THE 17TH CENTURY, THE JESUIT, GIANBATTISTA RICCIOLI MADE A CONSIDERABLE CONTRIBUTION TO THE NOMENCLATURE OF LUNAR FEATURES. HIS 'SELENOGRAPHIC MAP', COPIED AND COLOURED IN THE LATER WORK BY DOPPELMAYER, MAY BE COMPARED WITH THAT OF HIS CONTEMPORARY, JOHANNES HEVELIUS.

'waxing', because its brightness increases up to Full Moon. During the two weeks that follow Full Moon, the Moon is 'waning' up to New Moon.

Each of the four main phases of the Moon is separated from the subsequent one by an angle of 90°. As the Moon orbits the Earth and goes through its cycle of phases, each phase appears on a terrestrial meridian of longitude at a specific time. The New Moon and the Sun rise, cross the meridian and set together. Full Moon rises as the Sun sets; in other words, the Moon follows the Sun by twelve hours. First Quarter crosses the meridian as the Sun sets. Last Quarter is on the meridian as the Sun rises. Last, or Third Quarter is, in fact 3/4 of the way round the sky, which means that it follows the Sun by 3/4 of a day, that is, 18 hours, or precedes it by six hours.

The Sun is also a natural astronomical clock that defines the year. The *tropical year* is the interval of time between two spring equinoxes; it is the seasonal year, on which calendars that were to be of practical use for agricultural work needed to be based. But solar years have the disadvantage of not being easily observable. In contrast, the phases of the Moon are extremely easy to see. The first calendars were therefore constructed based on the movement of the Moon. In fact, although none of the numerous superstitions about the Moon and its possible physical effects on humans has any basis in fact, the Moon has at least played a major historical rôle through its effects on the development of calendars.

For practical reasons, a calendar month needs to contain a whole number of days, which is not the case with the lunation (approximately 29.5 days). A choice that consisted of reckoning a year as containing six months of 29 days, and six months of 30 days, would create a deficit relative to the true solar year. A purely lunar calendar cannot be synchronized with the seasons. To establish accurate astronomical calendars, people sought arithmetical combinations between the phases of the Moon and the cycle of the seasons, i.e., between lunar months and solar years. However, the year of 365.242 days, and the lunation, 29.350 days, are not commensurable. It is not, therefore, possible to find a rigorous combination that allows the motions of the Sun and the Moon to agree from time to time.

By making certain approximations, however, agreement was obtained, giving rise to what are known by the generic term of 'cycles': the *cycle of phases* repeats the lunar phases after a whole number of solar years.

A 19-year cycle, at the end of which the Sun and Moon return to almost exactly the same relative position, was known in China from the 22nd century BC. It was found by the Greeks in 432 BC. It is attributed to Meton; when the latter announced his discovery during the Olympic Games, the Athenians were so enthusiastic that they had the equation inscribed in the *agora* (the public square) in letters of gold, whence the name 'golden number' for the associated method of calculating when phases recur on a specific calendar date. The cycle (the *Metonic cycle*) is based on 235 lunations, or 6940 days, which are equivalent to 254 sidereal months, and 19 solar days of 365 5/19.

In 1900, Greek sponge divers discovered the wreck of an ancient vessel off the shore of the island of Antikythera. They brought up various treasures, which were exhibited at the archaeological museum in Athens. Among the booty was a compact block of calcified bronze and wood, the size of a large

■ This series of photographs of our satellite, taken at two-day intervals, show the change in the Moon's phases

bound book. In drying, the wood contracted, the concretion broke up, and the block opened up... The interior revealed traces of toothed wheels, incredibly complex gearing, and multiple inscriptions of an astronomical nature, dating back to the 1st century BC. The mysterious object was immediately dubbed the 'Antikythera mechanism'. Its secret was revealed in 1973 by a specialist in the history of scientific instruments, Derek de Solla Price. He realised that the gear ratios corresponded to a Metonic cycle of 19 years, during which the Moon completes 235 phase cycles and 254 sidereal cycles. The Antikythera mechanism was an elaborate astronomical calendar, enabling the prediction of the various lunisolar cycles, where the results were read off through the movement of a pointer on a dial – in some respects it was the ancestor of the famous movable card diagrams by Apianus!

But the golden number, despite its name, was (together with the Metonic cycle) not perfect: 235 lunations are actually 6939.688 days, and 19 tropical years are 6939.598 days, so that after one cycle, the lunisolar calendar is seven hours slow relative to the Moon, and nine hours relative to the Sun.

In 330 BC, Calippus obtained a closer approximation to harmonizing lunar and solar years. Assuming that the discrepancy in the Metonic cycle was exactly a quarter of a day, he multiplied the cycle by four, which gave a period of 76 years, or 940 lunations. Over this period, he lost one day. At the end of this cycle, the calendar was in agreement with the Sun, but was still five hours slow of the Moon. To remove this second discrepancy, Hipparchos subtracted one day every four Callipic cycles, i.e., 304 years, or 3760 lunations. Later he pushed the approximation to 4267 lunations.

The anomalies in the lunar orbit

The average distance of the Moon from the Earth is 384 400 km. Its orbit is actually elliptical, with a minimum distance, at perigee, of 356 375 km, and a maximum distance, at apogee, of 406 720 km.

The Moon's orbital period around the Earth is defined by the time that separates two successive passes at perigee. This cycle equals 27.55455 days and is known as the *anomalistic month*. In fact, the Greek astronomer Ptolemy, who did not know of the

OVER THE COURSE OF A LUNATION, THE PERIOD OF 29.5 DAYS THAT SEPARATES TWO SUCCESSIVE NEW MOONS.

elliptical nature of the lunar orbit, discovered that, during the course of a lunation, the Moon sometimes accelerated, and sometimes slowed down, which he considered as an anomaly.

The Moon's orbit is inclined 5° from the plane of the Earth's orbit, the ecliptic. At the time when the Earth was believed to be immobile in the centre of the universe, the ecliptic was defined as the apparent annual path of the Sun around the zodiac, but the two definitions are geometrically equivalent. The term 'ecliptic' comes from the Greek *ekleiptikos*, 'related to eclipses'. In fact, eclipses can occur only near points, called 'nodes', where the plane of the lunar orbit intersects that of the Earth, hence the name given to the latter plane.

Around its orbit, the Moon always presents the same face to the Earth: when following the progression of a cycle of phases, we always see the same craters. One might conclude from this that the Moon does not rotate. In reality, it rotates on its axis in a time equal to the time it takes to revolve around the Earth. Although we always see the same face, slightly more than half (59%) of the surface is actually visible over the course of a period of time. This is because of an slight apparent rocking motion of the Moon (both from side to side and up and down), known as *libration*.

The motion of the Moon is, of all movements in the sky, the most difficult to represent. The average angular velocity of our satellite is affected by several 'inequalities', which reveal themselves as a difference between predictions and observations. The most important had already been determined numerically by Hipparchos. Another anomaly, known as *evection*, was discovered in the year 138 by Ptolemy. Previous astronomers had recorded the position of the Moon at Full and New Moon only. By following the Moon around the whole of its orbit, he detected monthly accelerations and decelerations, which sometimes took it ahead, and sometimes behind, the positions calculated for the various quarters.

Other anomalies were detected by the accurate observations by Tycho Brahe. In his marvellous observatory sited on the Danish island of Hven, Tycho proceeded meticulously to measure the dates and durations of eclipses, to see if they agreed with the new theory of the motion of the Moon that he

■ This pastel drawing by E.L. Trouvelot, made in 1875, details the lunar relief around Mare Humorum. The craters are remnants of meteoritic impacts.

had just developed. On 28 December 1590, he calculated that an eclipse would start two days later at 18h 28m. The eclipse certainly started on the day in question, but while Tycho was still at table having his dinner! By the time he had assembled his assistants, the eclipse was almost over at 18h 05m. Furious and annoyed, Tycho momentarily abandoned observing the Moon! His sense of precision prompted him to resume his study some years later and he discovered several anomalies that had escaped his predecessors. He noted, for example, that the inclination of the lunar orbit is not constant, but varies between 4° 58′ 30″ at times of Full and New Moon, and 5° 17′ 30″ at the Quarters. He noted that the Moon accelerates in summer and slows down in winter. These anomalies were later explained by the mathematical genius of Johannes Kepler, who realised that in winter, when the Earth is closest to the Sun, the latter exerts a retarding influence on the Moon.

■ The same lunar landscape, photographed through the telescope a century later, shows the extraordinary quality of Trouvelot's work and of the great selenographers of past centuries, who combined manual dexterity with acute eyesight.

By dint of innumerable calculations, Kepler discovered that the orbits of the planets around the Sun and that of the Moon around the Earth are not circles, or compounds of circles, as had been thought since ancient times, but were ellipses. This enabled him to calculate new lunar tables, which were published in 1627 and used for several decades.

In addition to these major anomalies, the true motion of the Moon cannot be established without taking a large number of lesser equations into account. The first theory accounting for these small variations was given

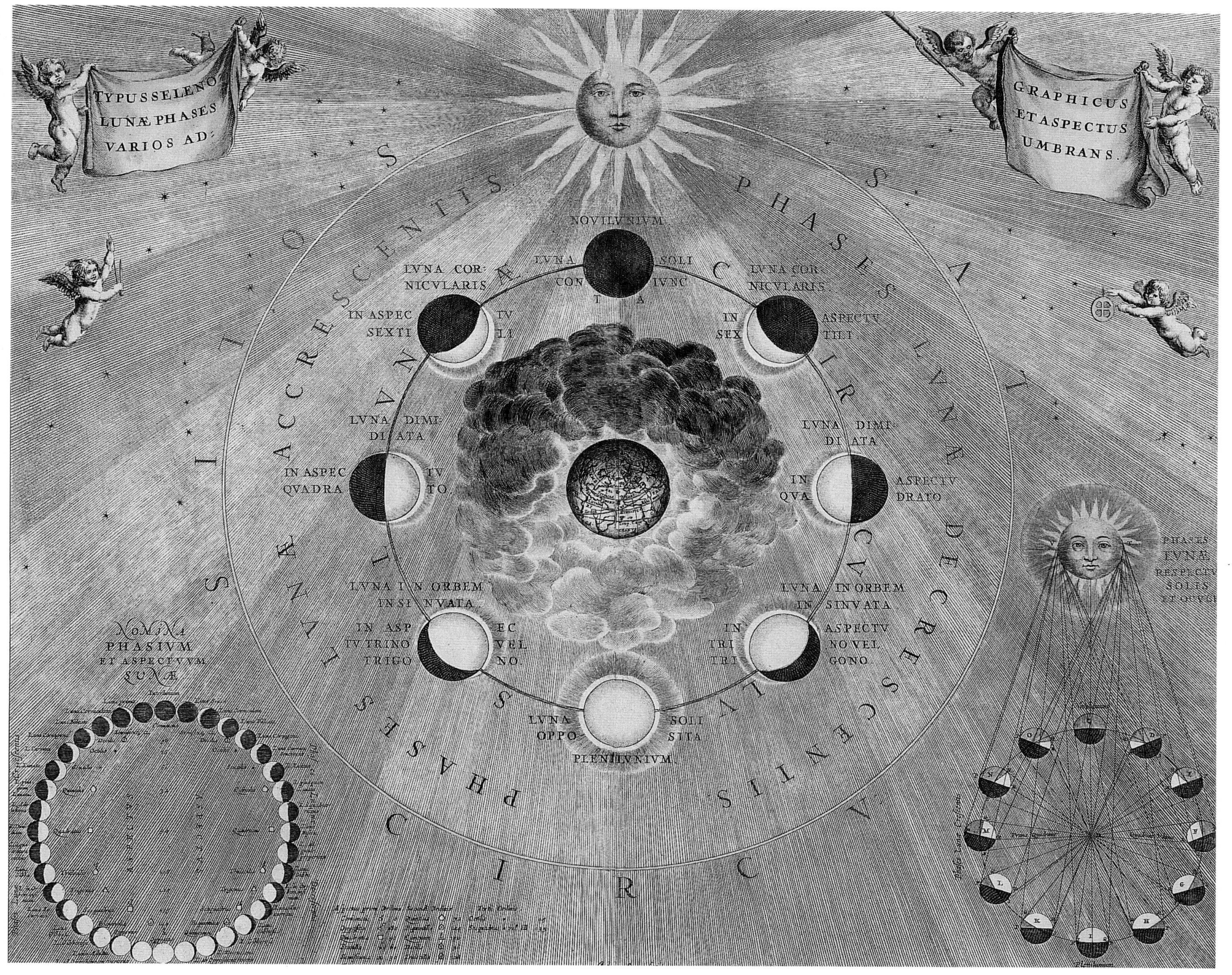

■ The cycle of lunar phases, determined by the relative positions of the Earth, Moon and Sun, is clearly explained in this luxurious 17th-century atlas compiled by Andreas Cellarius, the *Harmonia Macrocosmica*.

by Isaac Newton, in the context of his law of universal gravitation. Because every massive body exerts a force of gravitational attraction, the movement of the Moon is also affected by all the planets in the Solar System. Newton once admitted that his study of the Moon was the only thing that had given him a headache! In fact, calculation of the perturbations caused by the other planets was beyond the scope of mathematics in his day. The second edition of his main work, the *Principia*, which appeared in 1713, explained only eight inequalities, and left a residual difference between theory and observations to be resolved.

Reducing this difference was the truly monumental task facing the mathematicians of the 18th and 19th centuries. Alexis Clairaut, Jean Le Rond d'Alembert, and Leonhard Euler took the lead. The first two were initially against the Newtonian law of gravity, whereas the third, more prudent, declared that he would be 'sorry to dethrone Newton'. An intense rivalry developed between the three men. Clairaut had to admit his error in 1749: the law of universal gravitation was correct, but to ensure that it agreed with observations, it was necessary to include terms that had previously been neglected. D'Alembert arrived at the same conclusion by mathematical methods that were even more elegant. As for Euler, appointed in the meantime to the Academy at St Petersburg, his theory of the Moon was a failure. He then had the malicious idea of awarding the academic prize for 1752 to the best article devoted to the motion of the Moon. Clairaut fell into the trap and submitted his essay, which allowed Euler to discover his mistakes. As for d'Alembert, he was suspicious: convinced that Euler would favour Clairaut for the prize, he did not take part in the competition, and published his article nine months before anyone else!

In his *Theory of the Moon*, published in 1772, Euler defined the lunar motion by means of about fifty equations, each representing a small variation. But in practice, calculation remained impossible. Like Newton, Euler had to admit defeat: 'Each time, during these forty years, that I have tried to deduce the theory and motion of the Moon on gravitational principles, I have encountered so many difficulties that I am now forced to give up my work and all my earlier research.'

At the beginning of the 19th century, Tobias Mayer, retained just 14 perturbations and was able to obtain results that did not differ from the true positions of the Moon by more than 1.5 minutes of arc – which corresponds to an uncertainty for a

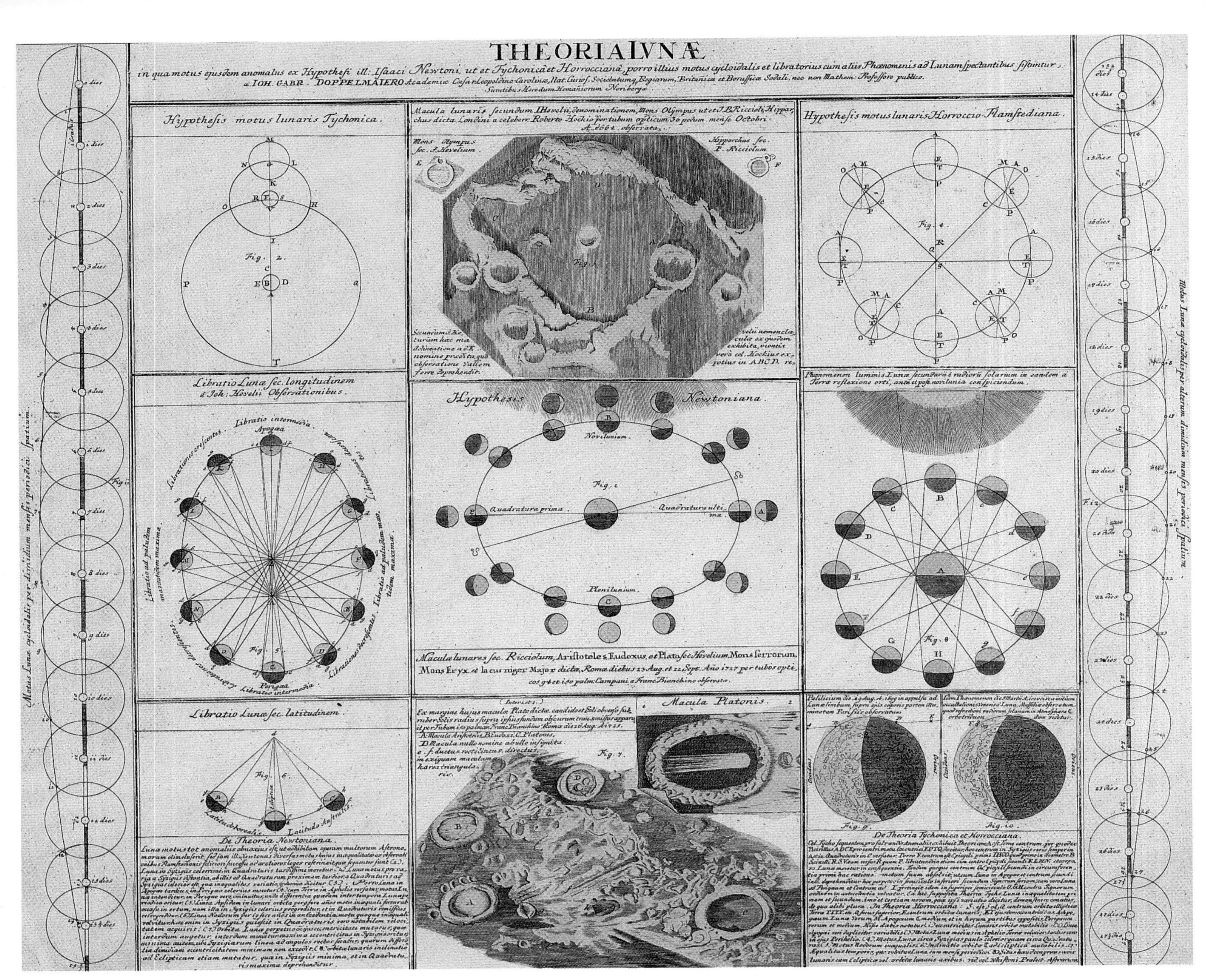

■ Theories about the Moon that aimed to explain the anomalies in its motion and to describe its relief, were developed, in particular during the 18th century. This plate comes from *Atlas Celeste*, compiled by Johann Gabriel Doppelmayer in 1742.

terrestrial observer of less than 100 km.

Joseph Louis Lagrange and Pierre Simon de Laplace, both brilliant mathematicians, also devoted a large part of their lives to lunar theory. Laplace succeeded in accounting for the motion of our satellite to an accuracy of 30 seconds of arc.

But the greatest exploit in this field was accomplished in the middle of the 19th century by an engineer, Charles Eugène Delaunay. He buckled down, on his own, to his great task and devoted twenty years to it: the first ten in developing his model, and the following ten to verifying his results. He published his conclusions in two colossal volumes, which appeared in 1860 and 1867. It was a century before anyone had the capacity to check Delaunay's work. It was the 1950s when the first electronic computers saw the light of day; at the same time as the first artificial-satellite launches, and the projects for lunar missions. In 1970, André Deprit and his collaborators ran Delaunay's model on their computer, completing in twenty hours what it had taken him twenty years to calculate. They found only three errors, two of which resulted from the first: a simple error in the transcription of a coefficient, which Delaunay had written as 33/16 instead of 23/16!

Nowadays, 1500 small lunar inequalities have been catalogued and taken into account into computer programs. They are needed to ensure the success of lunar missions.

The mechanics of eclipses

Eclipses of the Moon and Sun have evoked the greatest interest since the very earliest times. Anaximander, a disciple of Thales, and leader of the Milesian school around the middle of the 6th century BC, developed the first theory about them. We possess nothing by him except a short fragment quoted by Aristotle, but according to one of his commentators, Aetius, he declared: 'The Moon is a circle nineteen times larger than the Earth, similar to a chariot wheel, where the felloes [the pieces that form the rim] are hollow and full of fire, like the circle of the Sun, but situated obliquely relative to the latter. There is just a single blowhole, like the tube of a bellows, the phases follow the revolution of the wheel. The light of the Moon is its own. An eclipse of the Moon occurs by the closure of the vent that is on the wheel.'

According to Anaximander, the Earth is in the form of a cylindrical column whose height is one third of its diameter, surrounded by air, and then by fire. The celestial bodies were similar to wheels of fire, and their visible light was only one part of the whole, described as a vent through which the fire escaped. Eclipses of the Sun and Moon occurred through the vents being blocked. The phases of the Moon arose from the same basic cause.

These naïve attempts at an explanation nevertheless marked progress towards taking an abstract view, and also in the observation of nature. The fact that eclipses of the Sun do not occur except at New Moon and those of the Moon only at Full Moon must have been noticed early on. The fact that eclipses of the Sun were caused by the passage of the Moon in front of its disk, and that eclipses of the Moon were caused by the shadow of the Earth was discovered later. At the time of Anaxagoras (5th century BC), to whom this explanation is generally attributed (see Chapter 2), the idea was so little accepted among the people of Athens, which was full of an atmosphere of religious fanaticism, that a complaint was lodged against 'all those who do not recognize the reality of divine things and who profess learning on the subject of celestial phenomena'.

The periodicity of eclipses

The fact that the phenomenon of eclipses is recurrent was known to the Mesopotamians. Because eclipses result from the combination of lunar and solar conjunctions, a knowledge of the lunar and solar cycles is indispensable for determining the recurrence period. The occurrence of an eclipse does not depend just on the relative positions of the Sun and Moon, which determine the cycle of phases – in which case the Metonic cycle would suffice to predict eclipses – it also depends on the Moon passing one of the nodes of its orbit, so that the three bodies are aligned. The prediction therefore involves the *draconitic* month, the period after which the Moon returns to one of the nodes, and which amounts to 27.21222 days. A period that reproduces a similar succession of eclipses at the end of a whole number of lunations is known as an ecliptic cycle.

In accordance with relationships established in antiquity, the Chaldean astronomers discovered that the same succession of eclipses recurs at the end of 223 lunations. This cycle, equal to 6585 days, is known as the *Saros*. In years, it is equal to 18 years 10 days or 18 years 11 days, depending on whether four or five leap years are included.

At the time of Alexander's conquests, when he was leading his armies into Chaldean territory, Callisthenes sent his uncle Aristotle a list of observations of eclipses that had been obtained since 1900 years earlier. This would take knowledge of the Saros back to the Babylonians at the end of the 3rd millennium BC. The tablets in the *mul APIN* series, dating from the 8th and 7th centuries BC, indeed contains systematic observations of eclipses.

The discovery of the Saros was definitely not made via calculations, but by the accumulation of observations about the dates of eclipses, recorded over the centuries. The Saros mainly applied to eclipses of the Moon, because eclipses of the Sun are visible only from small areas of the Earth's surface, which are not the same at the end of a cycle, so that it is very difficult to establish their recurrence when remaining at a single site.

The discovery of the ecliptic cycle was not restricted to the West. The ancient Maya knew about lunar cycles, at least about the 19-year Metonic cycle. The *Dresden Codex*, a precious manuscript full of astronomical data, dating from the 12th century, is a copy, brought up to date, of a document from the 'Classic period' between the 4th and 9th century of our era. It contains elaborate calendars of the Moon and Venus. As regards eclipses, it included a table of dates extending over more than 32 years, and grouping together 405 consecutive lunations. These 405 lunations are divided into 69 groups, some of five, and others of six lunations, each group ending on a possible solar-eclipse date.

Calculation of eclipses of the Moon presents less complications than that of an eclipse of the Sun. In addition, the ancient astronomers, who were far from knowing the motion of the Moon accurately, had no means of predicting exactly eclipses of the Sun at a given spot. In fact, predicting the visibility of an eclipse at a particular point on the Earth assumes a knowledge of the distances from the Earth to the Sun, and to the Moon, and their relative sizes, of which the Babylonians were completely ignorant. The sole prediction that they could make was therefore whether an eclipse of the Sun was possible or not.

The reliability of the Saros for predicting eclipses is the result of remarkable arithmetical coincidences between the different lunar cycles. Eclipses recur at the end of 223 lunations because 242 draconitic months equal 6585.37 days, while 223 lunations equal 6585.32 days. This is the reason the Saros is reliable. Well, almost reliable, because like the Metonic cycle that governs the phases, the Saros is imperfect. We

■ This delightfully naive woodcut, from the *Merveilles de la Nature (Marvels of Nature)* by Conrad Lycosthenes (16th century), suggests the complex dance that the Moon and the Sun occasionally undergo in the sky.

■ These two views of the Moon during a partial eclipse were drawn by Amédée Guillemin for *Le Ciel* a famous popular-astronomy book of the second half of the 19th century, which rivalled that by Camille Flammarion.

have seen that innumerable anomalies perturb the motion of the Moon, of its nodes, and of the Sun. Normally these would have destroyed the fine regularity of the Saros. But an additional coincidence means that 239 anomalistic months – which, let us recall, measure the frequency with which the Moon passes perigee – equal 6585.54 days. This means that four points, and not just three, return to the same relative position at the end of 223 lunations, 242 draconitic months, and 239 anomalistic months! This remarkable arithmetic means that the Saros causes eclipses to repeat under almost identical conditions, although always with a shift of their region of visibility by 120° towards the west. It is easy to understand this difference. The precise value of the Saros is 18 years 11 days 8 hours, and not strictly speaking 18 years 11 days. A 360-degree rotation of the Earth takes 24 hours; eight hours additional rotation therefore corresponds exactly with a third of one terrestrial rotation, i.e., 120° longitude towards the west.

Kepler was the first to describe a satisfactory theory of solar eclipses in his *Astronomiae Pars Optica* (the *Optical Part of Astronomy*) of 1604, and in his *Rudolphine Tables* of 1627, he also gave a graphical method of determining where an eclipse of the Sun would be seen.

This new understanding of the mechanism of eclipses helped Christianity to establish its influence in the Far East. In the 17th century, western astronomy started to outstrip Chinese astronomy. The Jesuit missionaries who travelled to China intended to demonstrate the intellectual superiority of the West in secular fields, such as those of the sciences, and subsequently use this advantage to religious ends. In particular, they made use of their superiority in calculating eclipses; so successfully that in 1644 it was a Jesuit astronomer of German origin, Schall von Bell, who received the highly coveted post of head of the imperial bureau of astronomy!

Thanks to these methods, predictions of lunar and solar eclipses could be made several years in advance, as well as retrospectively. In 1770, the French astronomer, Alexandre-Guy Pingré published a *Chronology of eclipses of the Sun and Moon from the origin of our era until 1900*. The document founding the Bureau des longitudes, the French institute of celestial mechanics responsible for calculating ephemerides, gives him credit in these words: 'Astronomy has brought order to the chaos of the ages; without it, several ancient writers would have been incomprehensible. We know how much help it has rendered to those engaged in the art of verifying dates, and how much daylight Pingré has thrown on the history of the chronology of eclipses, based on the unvarying order of the motion of celestial bodies.' We shall see at the end of this Chapter how a certain form of chaos has returned to celestial mechanics... The fact remains that astronomers at that time were constantly trying to improve Kepler's method, along with progress in lunar theory. Friedrich von Bessel, in 1842, published the basic equations for calculating eclipses. Oppolzer was able to compute them with considerable accuracy from 1207 to 2161.

Nowadays we have extremely precise methods of predicting the dates and places at which eclipse will occur. Scientists have calculated the motions of the Earth and Moon so accurately that they can predict the occurrence of an eclipse to a few seconds, and define its region of visibility to a few tens of metres. The Internet sites of the Bureau des longitudes and NASA provide downloadable prediction charts. Astronomical programs that run on a portable computer allow anyone to prepare animated images. By choosing an appropriate integration time, one can witness, as if suspended in space, the motion of the Moon's shadow across the surface of the Earth.

■ The various colorations of the Moon during eclipses, drawn by Lucien Rudaux for his book *Le Ciel*, which appeared in 1923. The lunar eclipses involved are those of 10 March 1895, 17 December 1899, and 11 April 1903.

Eclipses of the Moon

When the Moon passes into the Earth's shadow, the light from the Sun is blocked. A lunar eclipse is visible from every point on the Earth where it is night-time. For obvious reasons of geometry, lunar eclipses cannot be seen except at the time of Full Moon.

The Earth casts a cone of shadow behind it, the boundaries of which are formed by the external tangents common to the bodies of both the Sun and the Earth. Any object contained within this cone (the *umbra*) does not receive any direct light from the Sun. A second cone, larger than the first, is formed by the interior tangents common to the Sun and the Earth. It is called the *penumbra*, and anything within this second cone receives light from just part of the Sun's disk.

A lunar eclipse is said to be 'central' when the Earth's shadow completely covers the Moon's disk, 'partial' when only part is covered, and 'penumbral' when the disk misses the umbra, but passes through the penumbra.

Because the shadow of the Earth is much larger than the diameter of the Moon at its distance of 384 000 km, the Moon may be darkened for as long as 1 hour 45 minutes in a central eclipse.

The Moon remains visible even during maximum eclipse, as an object with rather indistinct outline, and with a colour that ranges between bright red to a coppery red. This phenomenon of a red Moon, which was so striking to ancient peoples (*see* Chapter 2), is because the Earth's atmosphere refracts some of the light from the Sun onto the lunar surface. As it passes through the terrestrial atmosphere, the refracted sunlight loses blue light relative to red light – because the shorter wavelengths of visible light (violet and blue) are scattered more strongly by the atmosphere than longer wavelengths (red and orange). Dust and smoke particles suspended in the atmosphere accentuate this effect: the more there are at the time of the eclipse, the more the bright blood-red colour covers the disk of the Moon.

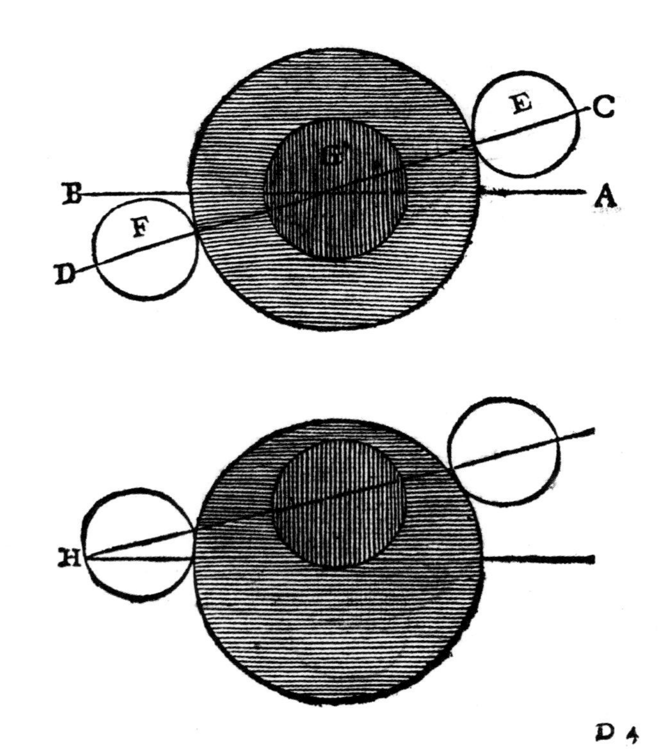

■ The *Oeuvres complètes* of the philosopher Pierre Gassendi include a volume that is entirely devoted to astronomy (1658). It give a clear geometrical description of eclipses of the Moon by the umbra and penumbra.

Eclipses of the Sun

The Earth orbits the Sun, and the Moon orbits the Earth in an anticlockwise direction (known as 'direct' rotation). As a result, the Moon sometimes passes between the Earth and Sun, and casts its shadow towards the Earth. An eclipse of the Sun necessarily occurs if, at the time of New Moon, the Sun is close to a node.

A lucky accident means that the Moon, which is roughly 1/400 times the size of the Sun, is also about 400 times closer to us, which means that it seems almost the same apparent size in the sky, hence the possibility that an eclipse may be total.

In fact, the distance between the Earth and the Sun changes between aphelion and perihelion in the Earth's orbit, so that the apparent diameter of the Sun varies between 31′ 31″ in early July and 32′ 35″ in early January. The distance of the Moon from the Earth also changes between apogee and perigee in the Moon's orbit, so that the apparent diameter of the Moon varies monthly between 29′ 22″ and 33′ 31″. Depending on the timing, the Moon therefore sometimes appears smaller, and sometimes larger than the Sun. Obviously a solar eclipse may only be total when the Moon's disk appears larger than that of the Sun.

When this is the case, the Moon's shadow reaches the Earth's surface, and people within the shadow find that all or part of the Sun's disk is hidden from sight. As for lunar eclipses, there are an umbral and a penumbral cone. The circular section of the umbra on the Earth's surface is much smaller than the diameter of the Moon, because the Sun is not at infinity. It is never more than 270 km across. But, as the umbra moves across the Earth's surface at a speed of approximately 1800 kph, the section of the umbra sweeps out a path that is several thousand kilometres long. This zone, called the path of totality, includes all the points for which the eclipse will be total. Despite its length, the narrowness of the path of totality explains why total eclipses are rare at any given point on the Earth.

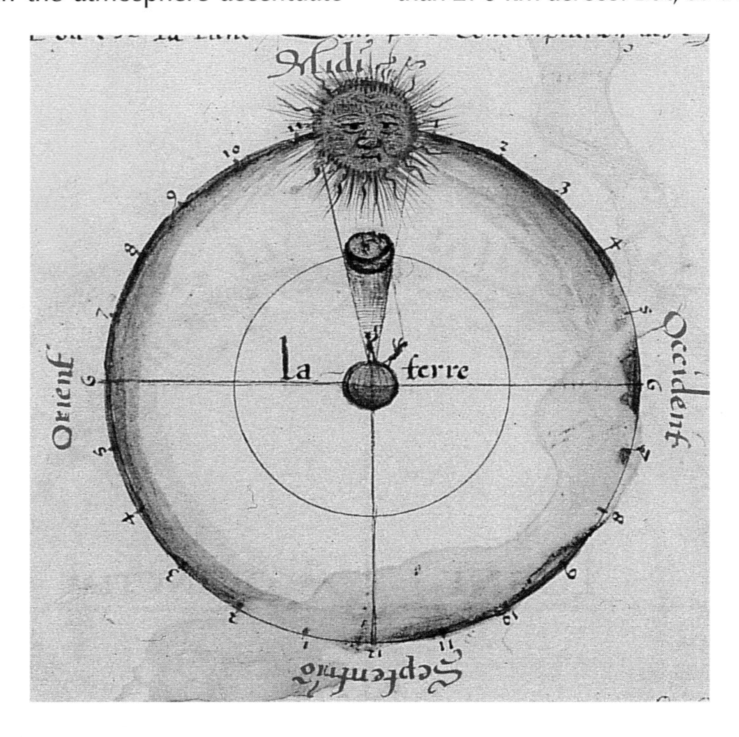

■ This 16th-century navigation manual by Jacques de Vaux contains astronomical instructions for the use of sailors. Here, the phenoenon of solar eclipses is explained, based on the geocentric system.

Outside the zone of totality, the eclipse is partial, and less and less of the Sun's disk is hidden the closer the observer approaches the edge of the penumbral cone. The maximum diameter of the section of the cone is about 7000 km.

The length of the Moon's shadow cone is, on average,

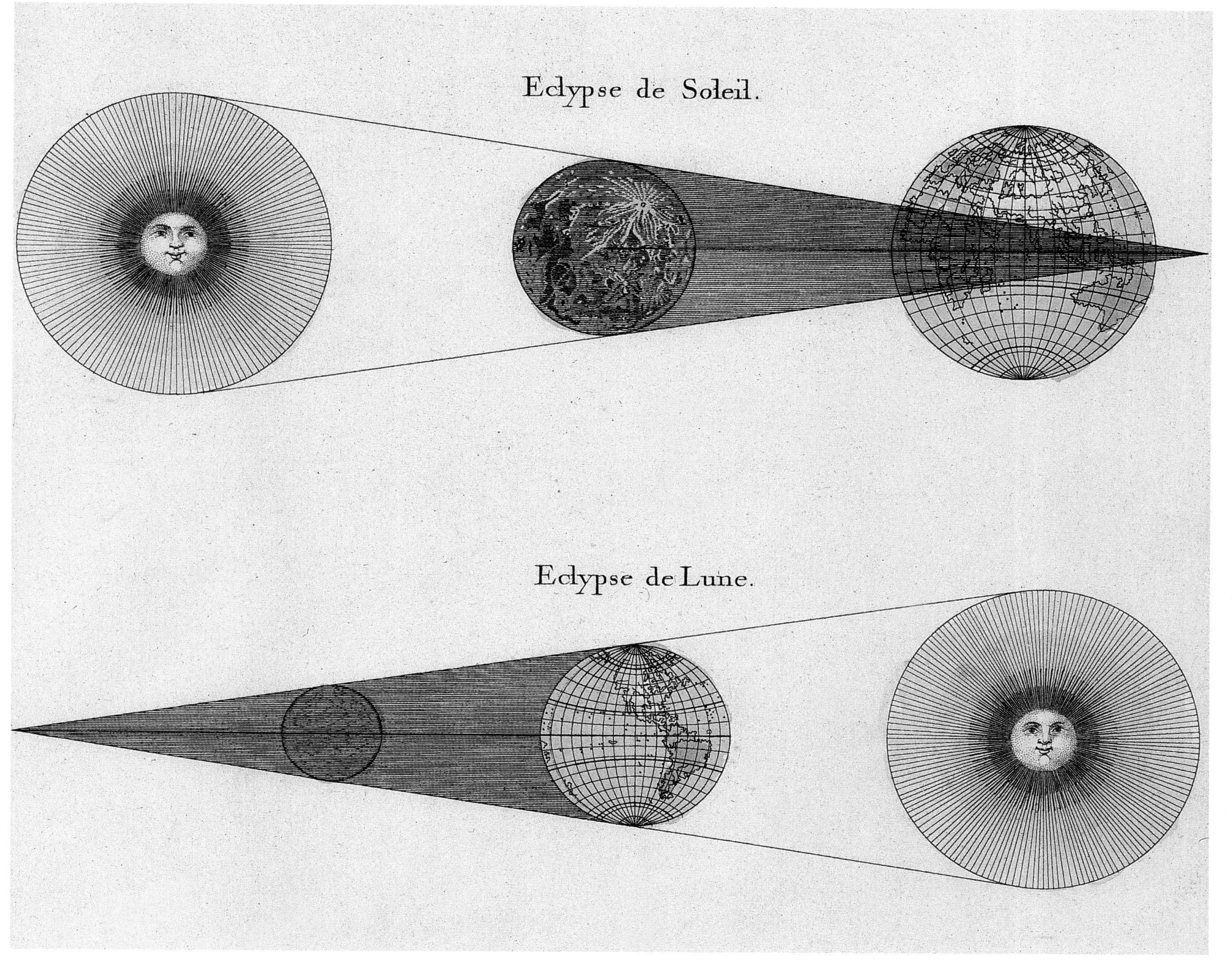

■ A lunar eclipse is visible from all points on the Earth where the Moon is above the horizon. By contrast, a solar eclipse is visible only from a relatively restricted zone. These diagrams are taken from an 18th-century atlas.

383 000 km. That means that if the Moon's phase reaches New at one of the nodes when it is close to its apogee (at a distance of 406 000 km from the Earth), the tip of the Moon's shadow cone will not reach the Earth. An observer situated on the axis of the cone will see a narrow ring of the Sun surrounding the black disk of the Moon. Such an eclipse is called an 'annular' eclipse. More than half of all solar eclipses are annular. It is a beautiful sight, but bears no comparison with that of a total eclipse, because the brightness of the ring of light prevents the corona from being seen.

The frequency of eclipses

In principle, in an ideal Solar System, where all the orbital planes were coincident, there would be two eclipses a month: one eclipse of the Sun at the time of New Moon, and one eclipse of the Moon at the instant of Full Moon. It is not like that, because, as we have seen, the Moon's orbit is inclined to the plane of the ecliptic and, for an eclipse to take place the line of the nodes (the intersection of the orbital planes of the Earth and the Moon), should point towards the Sun. Such favourable conditions are known as an 'eclipse season'.

The line of nodes oscillates by 19° each year around the centre of the Earth. As a result, the position of the nodes is not the same from one year to the next. In addition, the line of nodes rotates around the centre of the Earth in the opposite direction to the Earth's rotation. The effect is that eclipse seasons occur each year. But the arrival of an eclipse season does not automatically guarantee an eclipse: the Moon's phase also needs to be appropriate. A solar eclipse, for example, cannot occur unless the phase is New at the time that the Moon is at one of its nodes, and the same applies for a lunar eclipse at Full Moon. In a typical year, the eclipse seasons allow two solar eclipses and two lunar eclipses. But if an eclipse season falls early in the year, in January, for example, there may be three solar eclipses and four lunar ones, or vice versa, if one counts penumbral eclipses of the Moon. Seven is thus the maximum number of eclipses that may occur in the course of a year, and four the minimum. (In the latter case, there are two of the Sun, and two penumbra eclipses of the Moon.) There is not necessarily a total solar eclipse every year. There is none in the year 2000, for example.

A Saros includes, on average, 86 eclipses, 43 of the Sun and 43 of the Moon, again if one includes penumbral eclipses of the

Moon. If one counts only total and partial lunar eclipses, a Saros cycle includes only 71 eclipses, of which 28 are of the Moon.

At any given point on the Earth's surface, total eclipses of the Sun are exceptionally rare. Celestial mechanics and statistics tell us that, on average, one eclipse occurs every 370 years. Between the years 600 and 2000, for example, there were just two solar eclipses visible from Paris; the first on 16 June 1406, and the second – the last to date – on 22 May 1724. The next will not be visible until the morning of 3 September 2081!

The year 1999 saw two eclipses of the Sun, one total, the other annular. It is easy to understand why. To recap, a solar eclipse occurs only when the line of nodes points towards the Sun. The change in its orientation with respect to the stars is slow, however; so there is a good chance that after this condition has been fulfilled, it will be again six lunations later. The Earth-Sun line will have completed half a turn, i.e., 180°, without the line of nodes having had sufficient time to turn through a significant angle.

In addition, although the eclipse of 11 August 1999 was total, that of 16 February was not. The reason lies in the variation in the Earth-Moon centre-to-centre distance between perigee and apogee. Another factor that is involved is the length of the Moon's shadow cone. It varies according to the distance of the Sun. The closer we are, the shorter the cone: it was scarcely more than 368 000 km at the beginning of 1999, when the Earth was close to perihelion, but it exceeded 380 000 km in August the same year, shortly after the Earth had passed aphelion. On 16 February, therefore, the eclipse was annular, because the apparent diameter of the Moon was slightly less than that of the Sun. For a few tens of seconds, the Sun was visible as an extremely thin ring from anywhere within a narrow band that ran right across Australia, from an area near Perth to the north of Queensland. By contrast, on 11 August 1999, the eclipse was total, and the Moon's shadow, much longer, swept across the northern hemisphere.

The duration of solar eclipses

A total solar eclipse has a maximum duration when three conditions are fulfilled:

(1) the Earth-Sun distance is as large as possible

(2) the Earth-Moon distance is as small as possible

(3) the observer is at the least possible distance from the Moon; in other words, the Moon is as close to the zenith as possible.

■ Used since antiquity, armilliary spheres reproduce the celestial sphere, and are particularly suited to explaining the mechanism of eclipses. This luxurious example, made in Paris in 1705, was destined for the rooms of the Dauphin at Versailles.

Condition 1 is satisfied in July, when the Earth is at aphelion in its orbit around the Sun.

Condition 2 is fulfilled once a month, when the Moon is at perigee in its orbit around the Earth.

Condition 3 may be met when the eclipse is visible from a position with a latitude of +23° (i.e. 23°N).

Under such conditions, the diameter of the shadow cone is 262 km, and its velocity on the ground is 600 m per second. By dividing the diameter by the velocity, we obtain the maximum absolute duration of a total eclipse is 7 minutes 30 seconds.

The duration of annular eclipses is calculated by the same type of argument. It reaches a maximum of 12 minutes 30 seconds. As for the overall duration of a solar eclipse – from the time the Moon's disk begins to bite into that of Sun, to the moment when it reveals it completely – that is over six hours.

Over the course of 70 centuries between the year 2999 BC and the year 4000 AD, the Earth will have experienced 16 628 eclipses. The longest of these total eclipses will be that of 16 July 2186, with a duration of 7 minutes 29 seconds. The longest annular eclipse occurred on 7 December 150, and lasted for 12 minutes 24 seconds. The 20th century has been exceptional as far as the duration of eclipses is concerned. Three eclipses have exceeded seven minutes: those of 8 June 1937 (7 min. 4 sec.), 20 June 1955 (7 min. 8 sec.) and 30 June 1973 (7 min. 4 sec.). We have to go back to 1098 and forward to 2150 to find other eclipses that exceeded seven minutes!

Tides and the disappearance of eclipses

Why speak of the tides in a book about eclipses? Because both phenomena are caused by the combined effects of the Moon-Earth-Sun trio, visual effects in the latter case, and physical ones in the former. In addition, the tides may have a long-term influence of the very existence of eclipses.

It was not until Newton that the true mechanism for the tides was understood: they basically result from the Earth's reaction to the gravitational attraction of the Moon. Because the water in the oceans is more fluid than the rock of the Earth's surface, the Moon's gravity causes a bulge in the oceans directed towards the Moon. The water on the side of

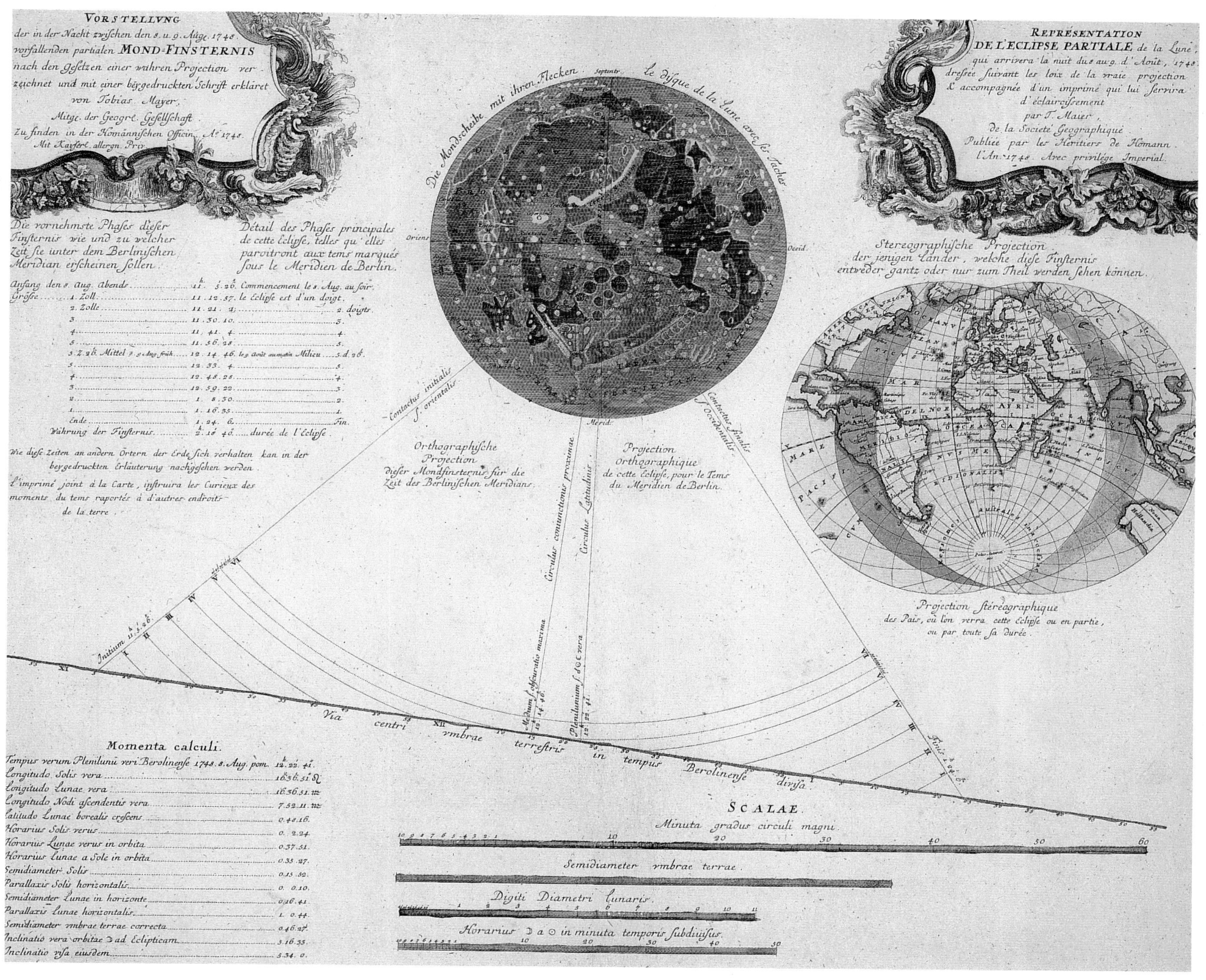

■ A representation of a partial lunar eclipse observed in 1748 by Tobias Mayer. The detailed map of the Moon indicates the regions of the lunar surface that were obscured or not by the Earth's shadow, whereas the terrestrial map shows the regions in which the eclipse was visible.

the Earth closest to the Moon experiences a greater force than that on the opposite side of the Earth. The difference between the amount of gravitational attraction produces a tidal force which creates two oceanic bulges. The continents are carried across each of these two oceanic bulges; the ground moves towards the water, which is deeper than normal. The water, in turn, is moving towards the land and produces a flood tide along the coast. When the Moon and the bulge have passed, the water flows away and the tide ebbs. Tides occur about 50 minutes later each day, because of the orbital motion of the Moon. This means that high and low tide are separated by approximately 6 hours 13 minutes.

The Sun also produces tidal forces, weaker than those of the Moon because of its great distance, but still significant, because of its extremely large mass. It is around the time of Full and New Moon that the range of the tides are greatest (spring tides), because the Moon and the Sun are working together to produce the greatest differential gravitational force. During First and Last Quarter, the Moon and the Sun are at right angles to one another. This means that their gravitational forces tend to compensate for one another. The range of the tides is then smallest (neap tides).

The tides create forces on both the Earth and the Moon which slow down their rotation. This phenomenon is known as tidal braking. This is what has locked one side of the Moon to face permanently towards the Earth, by reducing our satellite's period of rotation until it has become synchonized with its orbital period around the Earth. The tides also produce a long-term deceleration of the Earth's rate of rotation. As the Earth rotates, friction between the water in the oceans and the solid sea floors pushes the bulge forward of a line joining the centres of the Earth and Moon. The Moon's gravity tends to pull the bulge backwards, attempting to align it with the same Earth-Moon line. This retarding force is transmitted by the oceans to the Earth, and slows down its rate of rotation, just as a brake slows down a wheel. Tidal braking thus increases the time that it takes for the Earth to rotate by 2.3 thousandths of a second per century. If this tendency continues, a time will arrive when a solar day equals a lunar month, each of the two periods being equivalent to 47 of our current days. The two bodies will then continually turn the same hemisphere towards one another.

A third important effect of the tides that directly affects eclipses, is the progressive increase in the distance of the Moon:

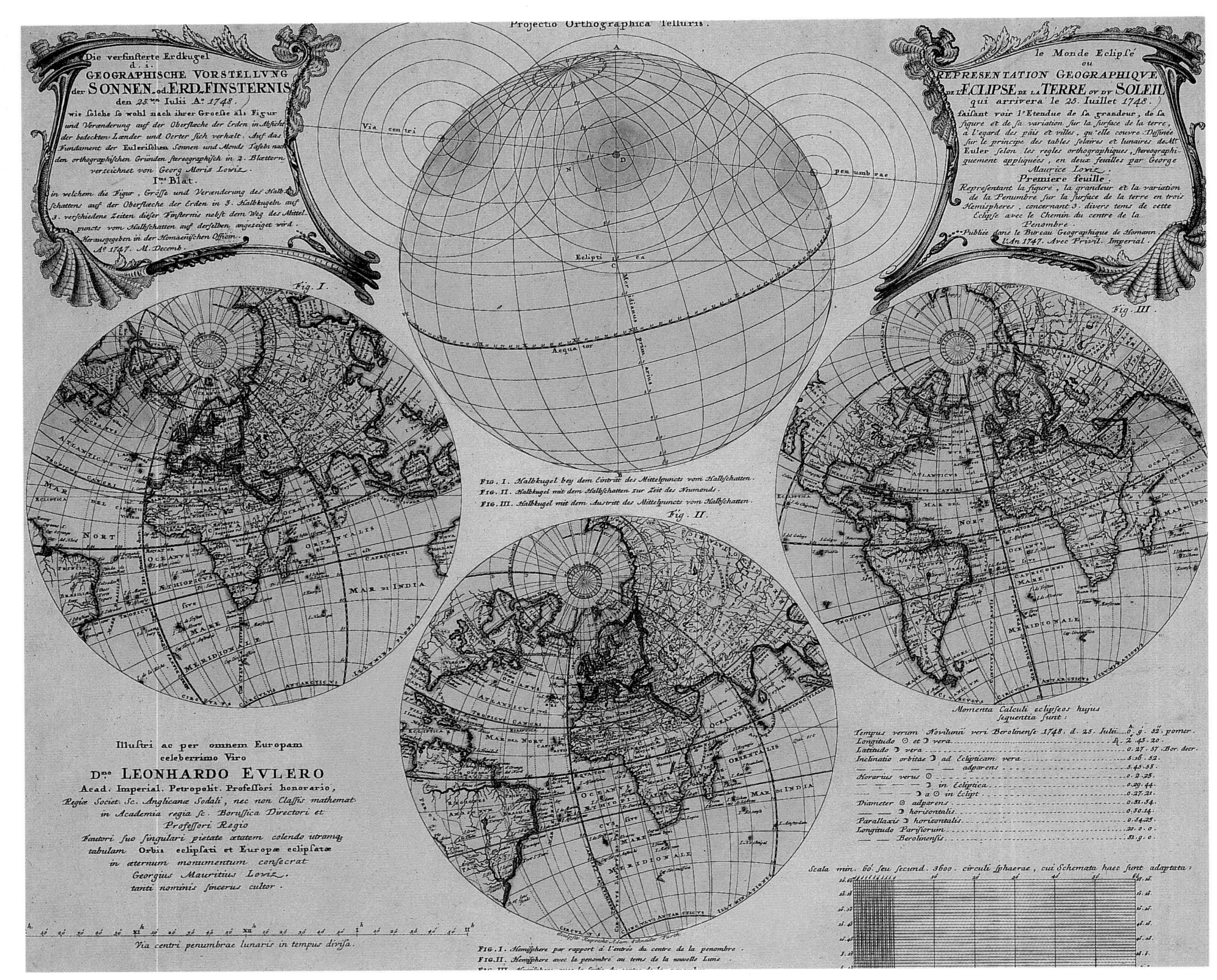

■ The mathematician Leonhard Euler, a professor at the St Petersburg Academy, developed solar and lunar tables which allowed him to predict extremely accurately the path of the eclipse of 25 July 1748.

our satellite is receding from the Earth in a slow spiral. To measure this drift, laser retroreflectors were installed on the Moon by the American Apollo and Soviet Luna missions. By sending pulses of laser light from the Earth to the reflectors on the Moon, scientists are able to determine the time required for the round trip, which gives the distance between the two bodies at any given moment to an accuracy of about 3 cm. Experiments of this sort are carried out on the Plateau de Calern in France at the Côte d'Azur Observatory. The results indicate that the Moon is currently receding from the Earth at a rate of 3.8 cm per year.

These measurements have also improved our knowledge of the Moon's orbit in the past, enough to allow precise analysis of solar eclipses back to 1400 BC.

If this tendency for the Moon to recede continues, we can imagine a distant time when total eclipses no longer occur. The apparent diameter of the Moon will have become permanently smaller that than of the Sun, so all eclipses will be annular ones. Extrapolating from current measurements, the 'last' total eclipse in the Earth's history will occur in some 200 million years, and will last for a brief second...

■ Nowadays, astronomical programs are able to calculate, in a few seconds, the dates, durations, and visibility of eclipses, as well as animated sequences of images.

Is this outcome inevitable? Nothing can be less certain! In actual fact, no astronomer can maintain that the tendency for the Moon to recede will continue for all eternity. The key factor here is *dynamical chaos*. This forbidding term hides a brilliant discovery made by the French mathematician Henri Poincaré at the beginning of the century, which was to revolutionize celestial mechanics. This theory, the scope of which remained unrecognized for fifty years, shows that the solutions of the equations of

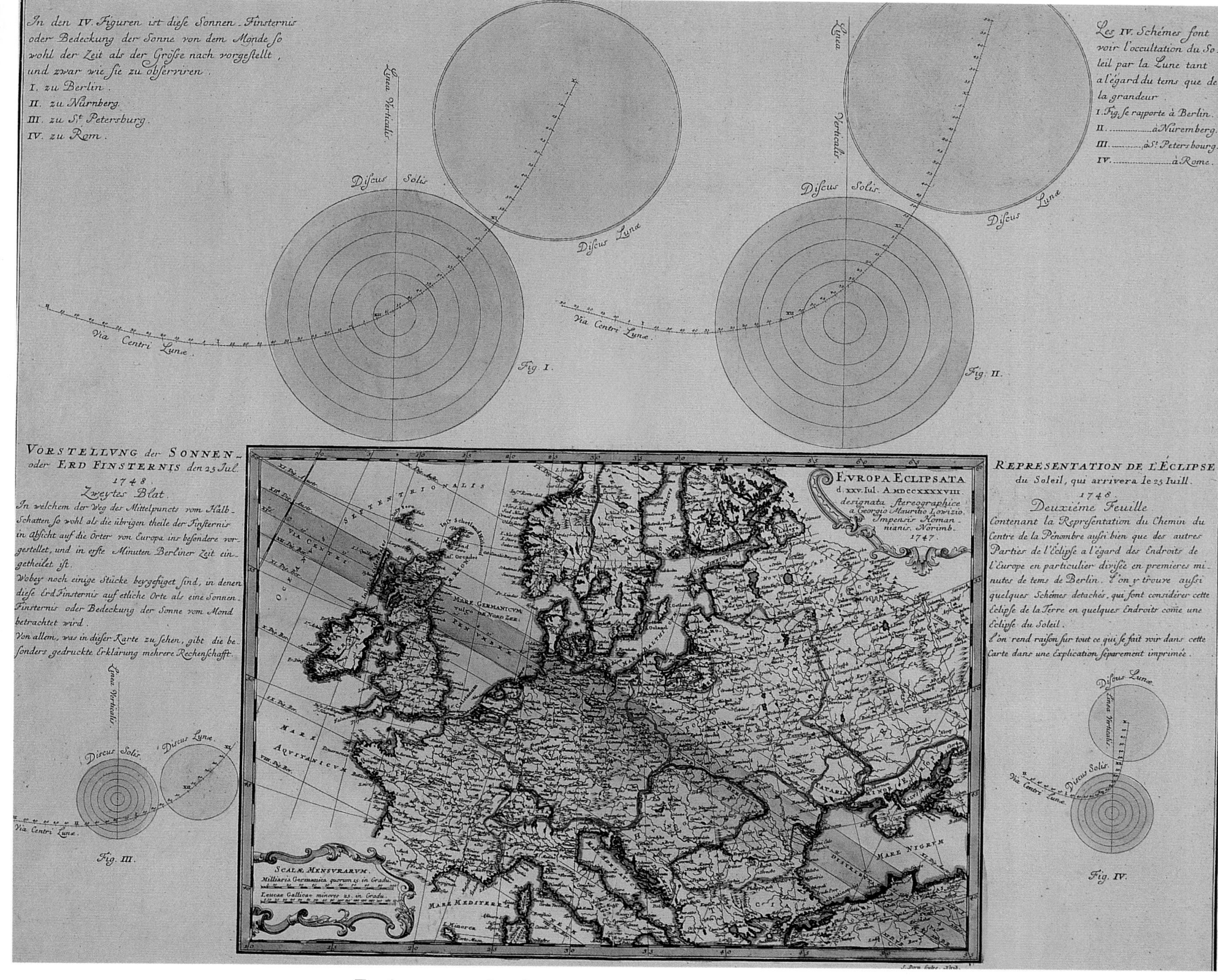

■ AGREEING WITH EULER'S CALCULATIONS, THE PATH OF TOTALITY OF THE ECLIPSE OF 25 JULY 1748 CROSSED SCOTLAND, NORTHERN GERMANY, BERLIN, ROMANIA, AND ON INTO TURKEY.

Newtonian gravitation include irregular motions that are absolutely unpredictable, as soon as more than two bodies are involved. In fact, such behaviour affects the mathematical description of almost all complex physical systems, so much so that one could say that the motion of the Moon is a benign case of a congenital disease: chaos.

As far as we are concerned here, the future evolution of tidal effects and the braking of the Earth's rotation depend on the combined gravitational influence of all the planets in the Solar System. As a result, it cannot be predicted over extremely long periods of time. It is as if there were an absolute horizon to mathematical prediction, regardless of the calculation power that is available.

The quest for order and predictability in celestial mechanics, so sought-after over the centuries, and of which eclipses have always been the most notable symbol, thus begins to experience false notes. But these discords let us glimpse a new and fascinating score, that only numerical simulation can unravel, without ever being able to reach a final goal. With dynamical chaos, it is the very fate of the Solar System that has become uncertain. But that is another story... ■

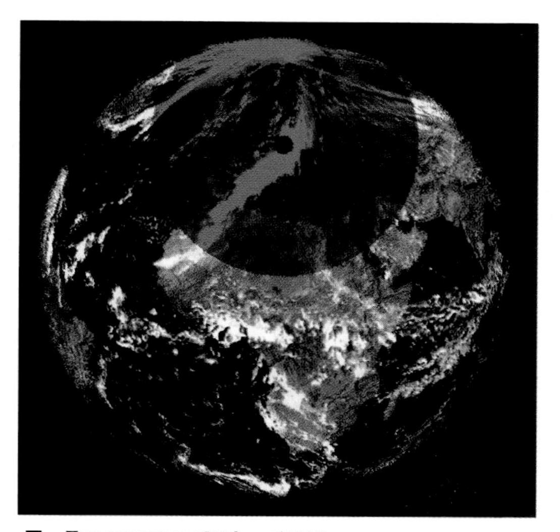

■ THE ECLIPSE OF 25 JULY 1748 AS RECALCULATED BY THE *REDSHIFT* PROGRAM (MAC OR PC), SEEN FROM THE MOON. THE TWO SNAPSHOTS SHOW THE PROJECTION OF THE SHADOW CONE OVER THE NORTH ATLANTIC (*LEFT*) AND PRUSSIA.

■ Eclipses and tides are two effects of the Earth-Moon pair. The latter, although still bound by gravity, is

slowly moving away, as a result of tidal effects. In the distant future, nowhere on Earth will experience a solar eclipse.

The great cosmic clockwork

■ At 5 o'clock, on the morning of 24 March 1998, at Mauna Kea Observatory on the island of Hawaii, dawn appears in a crystal-clear sky in which a few stars still glimmer. The moon is about to eclipse Venus, as the astronomers finish their night of observation, in the dome of the 3.6-m Canada-France-Hawaii Telescope. On the right, the first rays of dawn touch the dome of Gemini North, a telescope 8 m in diameter.

■ In the photograph of a fine occultation of Venus by the Moon, the part of the lunar disk that is in night-time darkness remains faintly illuminated by Earthshine.

Seen from a distance, one would say that it was a vast timepiece, with a small blue ball beating out the regular rhythm of the years, and around it, like some tiny second hand, the Moon marking the months. Lying farther away from the system's rotational axis, symbolized by the blinding Sun, there are other, more slowly turning gears, and their planets require decades, or even centuries, to complete a single turn around the dial. If one could speed up time, and see the planets and all their satellites moving at top speed, enough to make one giddy, one would first note the extraordinary regularity of this marvellous cosmic clockwork, then its great complexity, and finally, its irregularities.

Because the Solar System does not simply go round in circles: far from it, most of its hundreds of planetary and satellite orbits are elliptical rather than circular, and each gearwheel is affected by the presence of the others – accelerating, decelerating, and sometimes even sticking as a result of the presence of a cosmic grain of sand – over the course of aeons.

Our planetary system, is, above all, eight planets that revolve around the Sun in the same direction and practically in the same plane. Plus a rather incongruous gear train, probably a late addition, the small Pluto-Charon pair, whose eccentric and inclined motion would offend any good celestial watchmaker. Then there are some sixty-odd satellites, most of which orbit in more or less perfect circles and in a single plane around their respective planets. Finally, there are the small elements, the asteroids and comets, which are counted in the thousands, or maybe millions, and most of which remain unknown to astronomers.

Cosmic perspective effects

All these bodies are, of course, illuminated by the Sun. And because the orbital planes of the planets and their satellites are mostly coincident, or at least intersect one another... yes, there are eclipses throughout the Solar System. Some of them occur several times a day, and others are extremely rare, occurring only once a century, or even less often. Astronomers divide all these geometrical alignment phenomena into three major categories: eclipses, occultations, and transits.

Naturally, historically – or at least until the end of the 16th century – eclipses involved only the Sun and the Moon. Then, with the Renaissance and the invention of the telescope, astronomers started to realize the universal nature of these perfect celestial alignments, and discovered that a number of satellites caused eclipses on their planets.

As far as occultations are concerned, they are merely one particular form of eclipse, where one body hides another, that has a far smaller apparent diameter: the Moon and the planets

■ The complete series of an occultation of Venus by the Moon, recorded in a single image by the Japanese photographer Akira Fuji. The Moon slowly approaches Venus before hiding it, while the apparent rotation of the sky inexorably carries both bodies towards the western horizon.

■ This image, taken during the total eclipse of 26 February 1998, records the Sun, Moon, and the planets Mercury (*left*) and Jupiter (*right*). Such a close approach during an eclipse is extremely rare.

are primarily involved.

Finally, transits are a form of mini-eclipses, or even micro-eclipses. These are transits of satellites in front of their planet, or of planets in front of the Sun, the latter being as seen from the Earth.

All these eclipses, to use the term broadly, passionately interest certain observers, who devote their free time, if amateurs, or their careers, if professionals, to seeking out some wonderful and rare arrangement that the great cosmic clockwork will create tomorrow, in a century's time, or in a thousand years. Some of these phenomena were, in the past, of extreme scientific interest; others owe their beauty simply to the fact that they are a hundred times as rare as a total eclipse of the Sun. These events are, in fact, mathematical and mental abstractions, either because, never having been observed in the past, no one really knows if they will indeed occur one day, or because in all probability they will never be seen by anyone: the improbable encounter between an eclipse and its admirers is set to occur in a thousand years, but on another planet, three billion kilometres away...

Just like eclipses of the Sun and Moon, occultations and transits may be partial. Again like eclipses, these short-duration phenomena cannot be observed from just anywhere: sometimes the region of the Earth from which they are visible is extremely limited. Moreover, outside this zone of visibility, the eclipse or the occultation will not occur, but there will still be a spectacular close approach, which astronomers call a 'conjunction'. Close conjunctions between the Moon and the brightest planets, Venus, Mars, Jupiter, and Saturn delight amateur astronomers. Frequently, at twilight, these planets and our satellite, aligned as if on parade in the sky, and pointing an invisible arrow towards the Sun, which has already set below the horizon, provide a fine lesson in celestial mechanics by reminding us of the main feature of the celestial clockwork: all these gears are more or less confined to a single plane.

In fact, during total eclipses of the Sun and Moon, a lucky

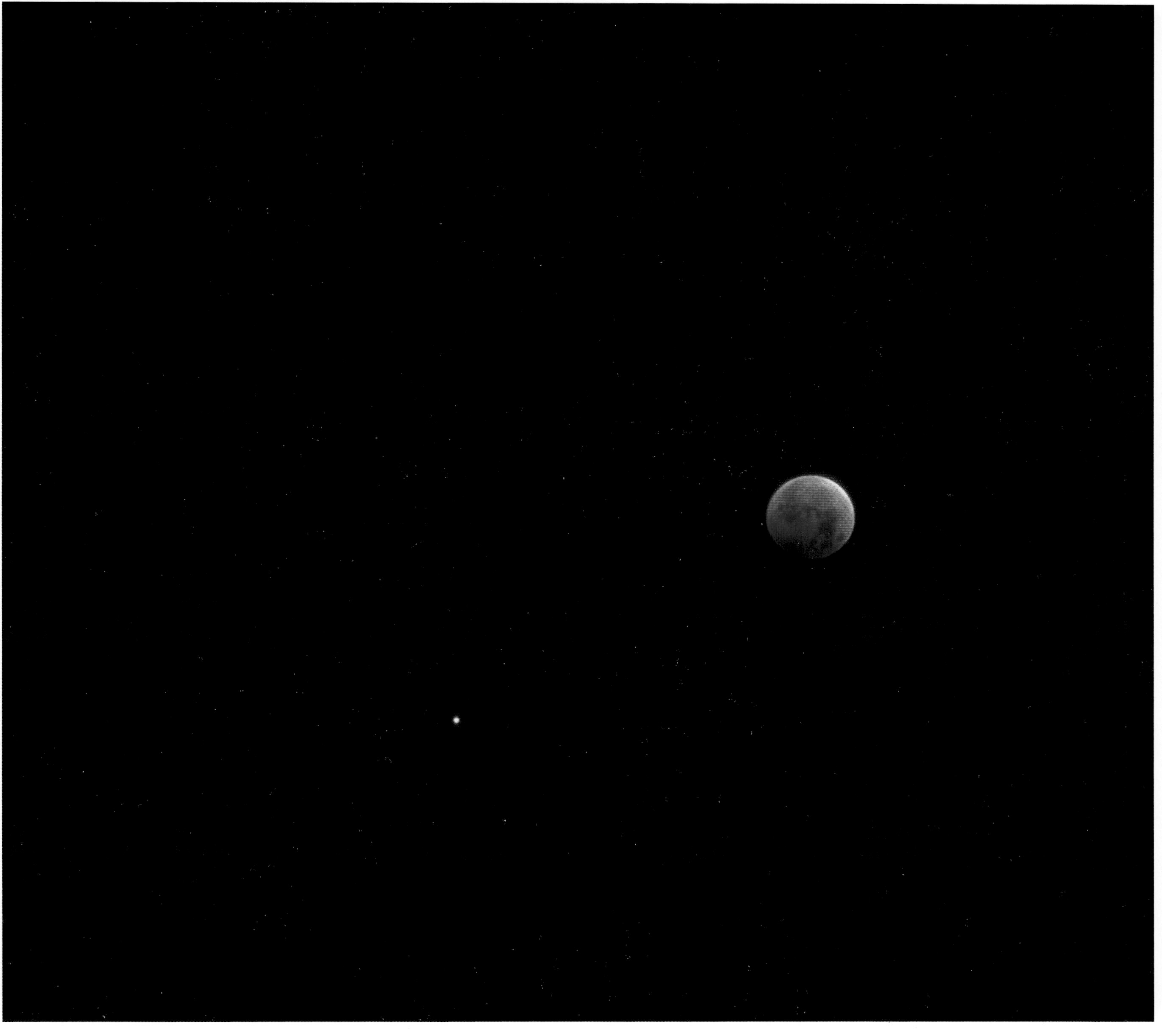

■ The surprising sight of the planet Saturn, more than one billion kilometres away, rivalling the brightness of the Full Moon. This image was obtained during the total lunar eclipse of 27 September 1996.

chance sometimes means that planets are close to the perfect alignment between the Sun, Earth and Moon. The total lunar eclipse of 27 September 1996, for example, saw a spectacular approach of the Moon and Saturn, which were just 2°20′ apart. The planet, initially drowned by the flood of light from the Full Moon – which, we may recall, has an apparent diameter, like that of the Sun, of about 0.5°, or 30′ – before the eclipse began, became more and more conspicuous as the Moon plunged into the Earth's shadow. At 3h 59m, towards the end of the night and at maximum eclipse, the dark red Moon was scarcely any brighter than Saturn: its brightness had fallen, during the eclipse, by a factor of ten thousand! On 3 November 1994, it was Venus that came within about 5° of the Sun during a total eclipse. Finally, on 26 February 1998, delighted observers admired a planetary conjunction that brought Jupiter, Mercury, and a total eclipse of the Sun together. On that day, in fact, the two planets were 2°42′ and 4°5′, respectively, away from our star. Such a sight will not, unfortunately, recur very soon. During the eclipse of 21 June 2001, Jupiter will be 5°3′ from the Sun; for the total eclipse of 1 August 2008, Mercury will be 3°25′ away; then 3°10′ during the eclipse of 14 December 2020, and 3°5′ during the eclipse of 4 December 2021. Above all, however, we will have to be patient until 2 August 2027 to see Venus exceptionally close to the Sun at just 2°48′.

These conjunctions between planets and their star, when it is momentarily dimmed, add considerably to the magic of eclipses. They bring a new dimension to the spectacle, as if they truly set the Solar System in perspective. Even rarer than eclipses themselves, these close conjunctions are, however, observed from time to time over the course of decades. On the other hand, to this very day, no one has ever seen the Moon occult a planet during a total eclipse! Such a phenomenon has probably occurred in the past, and will certainly occur in future... it is, however, impossible to predict. The equations of celestial mechanics, despite their extraordinary precision, are powerless to cope with the vast number of variables that need to be taken into account in such a calculation. With irregular variations in the Earth's rate of rotation, cyclical slowing down and recession of the Moon (all unpredictable), and uncertain gravitational perturbations between planets: we just have to accept that beyond a few thousand years, the future of the Solar System is indeterminate.

As seen from Earth, of course, it is the Moon that is most likely to meet other bodies along its path. In a month, in fact, our satellite completes a circuit of the whole sky. Its path, which oscillates from one side to the other of the ecliptic – the Sun's apparent path in the sky, or, which amounts to the same thing, the projection onto the sky of the plane of the Earth's orbit – is never perfectly the same. Some planetary occultations are never observed. Those by Mercury, the closest planet to the Sun, always take place in full daylight, and are unseen by any telescopes. Those by Uranus and Neptune, two distant planets, invisible to the naked eye, are extremely difficult to see, and require the use of powerful instruments. To set this in perspective: the Full Moon is one hundred million times as bright as Neptune.

A few planets, in contrast, although they are rarely occulted by the Moon, do fare fairly well with respect to brightness. Occultations of Venus, Jupiter and Saturn offer observers a magnificent sight, despite the difference in brightness: the Moon is about 250 times as bright as Venus, and ten thousand times as bright as Saturn. It is an unforgettable experience to see, through a telescope, the immense lunar disk, often partially in darkness, slowly approach the giant planet, which looks like a minute yellow ball encircled by a white ring. In the field of a telescope, Saturn, like Jupiter and Venus, is about one-fiftieth of the size of the Moon, and when looking though the eyepiece, the perspective effect is striking and almost surreal. In fact, any large lunar crater appears larger than the planet,

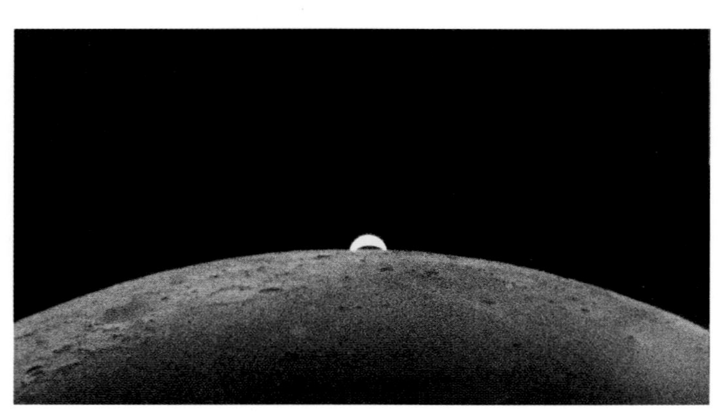

■ Observed through a telescope on the morning of 12 July 1996, this beautiful occultation of Venus by the Moon clearly displays the brilliant crescent of the planet. Venus, covered in a bright layer of clouds, reflects light far better than the dust- and lava-covered plains of the Moon.

which, with its rings, does actually extend for nearly 300 000 km! It is true that, although the average distance of the Moon is 380 000 km, that of Saturn is around 1.4 billion kilometres. When an occultation occurs, it takes our satellite more than a minute to hide the disk of the planet. About an hour later – i.e., the time that it requires for the Moon to cover a path in the sky that it equal to its own diameter – it slowly reveals the hidden planet. Occasionally, the Moon and the planet merely brush past one another. These 'failed' occultations are even more astonishing: for several minutes one can actually see the planet be partly hidden by a lunar mountain, only to reappear farther on, then be hidden by the rampart of a crater, before slowly moving away from the Moon. And then, much, much more rarely, a pair of planets are found together in the path of the Moon. Such an event occurred on 23 April 1998. Very few observers saw it, because to do so meant being, in the early hours of the morning, in the middle of the South Atlantic. Only a few lucky ones, most of whom were dedicated eclipse chasers, travelled to Ascension Island and St Helena. There, they were able to admire the sight as an attractive crescent Moon occulted, at an interval of forty minutes, Jupiter and Venus!

Although encounters between the Moon and planets are rare, each month, in contrast, our satellite occults thousands of stars. Obviously, most of these events remain invisible, because the extreme brightness of the crescent or even greater phase of the Moon, 'eclipses' – if we dare use the term – the brightness of all the stars in its vicinity. Some stars, however, are sufficiently bright to remain visible, despite the brilliance of the Moon. This is the case, for example, with the Pleiades cluster (a group of stars that is regularly occulted by the Moon), of Regulus in Leo, or Spica in Virgo; stars that are some of the brightest in the sky and which, before being eclipsed, form a fine pair with the Moon.

Venus, Jupiter, Aldebaran, Antares

Such stellar occultations remain, even as we enter the 21st

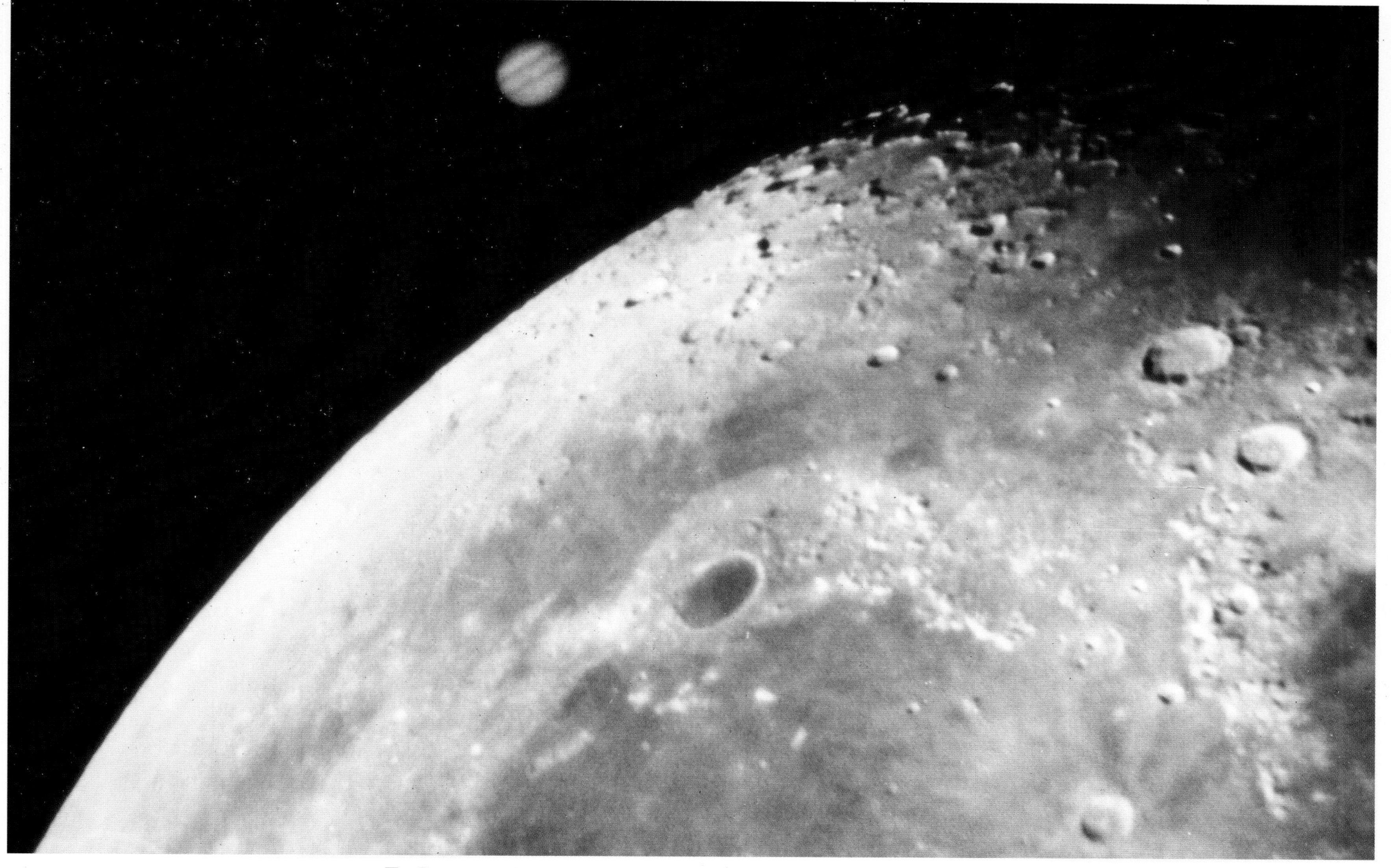

■ PLANETARY CONJUNCTIONS OFFER STRIKING EFFECTS OF COSMIC PERSPECTIVE. HERE, THE MOON IS ABOUT TO OCCULT JUPITER, THE LARGEST PLANET IN THE SOLAR SYSTEM. IN THE FOREGROUND WE CAN MAKE OUT MARE IMBRIUM AND THE DARK FLOOR OF PLATO.

century, of great interest to astronomers. In fact, the coordinates of stars on the celestial sphere are known to a high degree of accuracy, in particular thanks to the European Space Agency Hipparcos mission, which measured the positions of more than 400 000 stars to an accuracy of 0.001″. But what about the Moon? This moves and is slowly receding from the Earth because of complex gravitational interactions with its parent planet, the Earth. Moreover, because it has a significant, but irregular, apparent diameter, its position in the sky, at any given moment, is not easy to determine. The stellar background gives astronomers an almost perfect coordinate grid, which they regularly use, during occultations, to determine exactly where the Moon is. Regulus or Spica, lying some tens of light-years away from Earth (i.e., hundreds of thousands of billions of kilometres), subtend an apparent diameter that is essentially zero: to observers they appear as points. When the Moon hides them, unlike planets, their light is extinguished almost instantaneously! The sight is particularly striking if the occultation occurs when the Moon is young and waxing. Under these conditions it is the dark portion of the Moon, where it is night-time, that hides the star. Through a telescope magnifying 200 or 300 times, one can then see Spica or Regulus gently scintillating in the night sky, and then, without warning, abruptly disappear. In the case of Aldebaran in Taurus or Antares, the brightest star in Scorpius, observations of occultations are even more fascinating. Antares is a red supergiant, ten thousand times the brightness of the Sun, lying at a distance of some 700 light-years. Its actual diameter is gigantic, around 500 million kilometres. As seen from Earth, the star appears as a minute disk, about 0.05″ across, too small to be detected as such in a telescope. The Moon, however, moves across the sky at a rate of 0.5″ per second. It therefore takes about one tenth of a second to occult Antares, and this non-instantaneous extinction is perceptible to a trained observer, as is that of Aldebaran, a red giant whose apparent diameter is almost as great as that of Antares.

More generally, by timing precisely the instant of immersion and emergence of a star behind the Moon's disk, researchers are able to better understand the motion of the Moon around the Earth. For astronomers, however, there are other ways to use occultations and eclipses. This involves searching for such events in the past, and comparing, by a sort of retrospective prediction, what past observers ought to have seen with what they really did see. Such research is full of information, as is shown by the analysis of the spectacular occultation of Aldebaran that was admired at Athens in the year 509. More than a thousand years later, in 1718, the famous astronomer Edmond Halley – the man who predicted the return of the comet that now bears his name – tried to work out, by calculation, the path of the Moon across the constellation of Taurus in 509, and realised to his amazement, that, in principle, the occultation could not have taken place, because the Moon would have passed well to the south of the bright star! Sure of the reality of this occultation, but also sure of the reliability of his own calculations, Halley announced that, if there really was an occultation, it was because Aldebaran, at that earlier time, was several minutes of arc from its position in the 18th century. The British astronomer was right: he had just discovered stellar

■ On the night of 30 November 1994, the Moon occulted Spica in Virgo. Despite their apparent proximity in the sky, the two bodies were separated by an incredible distance. The star, lying at a distance of about 300 light-years, is actually ten billion times as far away from us as the Moon.

proper motion. In Aldebaran's case, at its distance of 65 light-years, this movement is of the order of 8′, i.e., a quarter of the diameter of the Moon, in two thousand years.

■ ALDEBARAN, A RED GIANT AT A DISTANCE OF 65 LIGHT-YEARS IN THE CONSTELLATION OF TAURUS, IS ABOUT TO DISAPPEAR BEHIND THE MOON'S DISK ON 14 MARCH 1997.

RESOLVING THE MYSTERY OF THE QUASARS

It was, however, another occultation of an object by the Moon that was to remain engraved in astronomers' memories. It took place on 5 August 1962. That was the day that, as seen from Parkes Observatory in Australia, 3C 273 would be occulted by the Moon. This extraordinarily powerful radio source, discovered in 1959, intrigued astronomers, who were quite unable to determine its size, distance or absolute magnitude. With a radio telescope, 3C 273 appeared as a large diffuse object. In the 1960s, one of the major problems for the young discipline of radio astronomy was the identification of the sources observed. Radio telescopes' Achilles heel is their poor angular resolution. The smallest perceptible detail of a celestial image is a function of the diameter of the telescope and of the wavelength at which observations are made. The longer the wavelength, the greater the diameter of the mirror should be. Radio telescopes observe at wavelengths of between 1 cm and 1 m. To reach the standard resolution of 1 arc-second that is obtained with classic optical telescopes, radio astronomers would have required an aerial 10–100 km in diameter! It is easy to understand why, observed by the Parkes 64-m aerial, 3C 273, appeared diffuse.

Cyril Hazard decided to take advantage of the object's occultation by the Moon. Given that the movement, position, and shape of our satellite are known to a high degree of accuracy, all that was required was to time the exact instants that 3C 273 disappeared and reappeared to be able to determine its exact position on the celestial sphere, as well as its size. On 5 August 1962, at 17h 46m, under the watchful eye of the Parkes radio telescope, the radio source suddenly disappeared behind the Moon. The astronomers Cyril Hazard and his colleague John Bolton, who were convinced that they had obtained an observation of crucial importance, duplicated the records of the occultation, and decided to take them to Sydney University on board two different aircraft – just in case...

In October 1962, the analysis of the Parkes recordings were finally finished. The astronomers now knew the position of 3C 273 to within 1 arc-second. Immediately, the optical counterpart of 3C 273 was found on photographic plates taken at Palomar Observatory. It was bluish star, from which a curious, linear luminous jet was escaping. In December 1962, the Dutch astronomer Maarten Schmidt turned the 5-m Mount Palomar telescope onto 3C 273 and recorded an incomprehensible spectrum. Six strong emission lines did not correspond with any known chemical element. Schmidt studied the spectrum of 3C 273 in vain for two months, before he realised, on 5 February 1963, that these unknown lines were simply those of hydrogen, but shifted by 16% towards the red – astronomers speak of a redshift of 0.16. By interpreting this shift in terms of a recession velocity, in accordance with the theory of an expanding universe, all that had to be done was to calculate the distance of the object, and, simultaneously, its intrinsic luminosity. The astronomers had just made an incredible discovery: 3C 273 was not a star, but a completely new type of object, lying at a distance of about two billion light-years, and emitting energy equivalent to a thousand galaxies similar to our own. The first quasar had been discovered, thanks to a lunar occultation.

Although the Moon regularly occults stars and planets, far more rarely, the planets occult one another! Specialists in orbital mechanics have calculated that twenty-two mutual planetary occultations have occurred, or will occur, between 1522 and 2223. None of these took place during the 20th century, the last occultation taking place on 3 January 1818, when Venus eclipsed Jupiter. In fact, it seems that just one eclipse of one planet by another has ever been observed telescopically throughout the whole history of astronomy. On 28 May 1727, from Greenwich Observatory, the astronomer John Bevis saw, through his telescope, Venus slowly passing in front of Mercury. In any case, before that date such phenomena were practically undetectable by astronomers: the first telescope was turned

■ A GRAZING OCCULTATION OF VENUS BY A NICE CRESCENT MOON ON THE EVENING OF 2 DECEMBER 1989. THE PLANET SEEMS TO SKIM THE MOUNTAINOUS PROFILE NEAR THE SOUTH POLE OF THE MOON.

■ It was an extremely rare event that was witnessed by the eclipse chaser Olivier Staiger from Ascension Island: the almost simultaneous occultation of two planets by the Moon...

towards the sky by Galileo in 1609, and Isaac Newton made the first reflecting telescope in 1668. Nowadays, hundreds of thousands of amateur astronomers around the world, dream of witnessing this strange and magnificent sight. Unfortunately, they need to be very patient: the next mutual planetary occultation should occur on 22 November 2065, when Venus passes in front of Jupiter. But the occultation that really stands out, requires an even longer wait. On 25 January 2518, Venus will pass precisely in front of Saturn. For a few minutes, our descendants will witness a sight that no one has ever seen: the planet Venus surrounded by rings!

Astronomers short of exotic eclipses, may, instead of waiting for the occultation of Saturn by Venus, witness those offered by Jupiter's entourage practically every day. The largest planet in the Solar System lies some 780 million kilometres from the Sun. It is accompanied by four large satellites, discovered by Galileo in 1609. The moons are a respectable size: Europa, the smallest is 3130 km in diameter, and Ganymede, the largest, 5280 km. By comparison, the Moon's diameter is 3476 km. Each time they pass between the Sun and their parent planet, Io, Europa, Ganymede and Callisto cause a total eclipse on Jupiter's surface. Or rather, on its clouds, because Jupiter is basically a gaseous world. Seen from Jupiter, the Sun is minute: its diameter does not exceed 6 arc-minutes, or one-fifth of the solar diameter as seen from Earth. As seen from Jupiter, Io, Europa and Ganymede have apparent dia-meters of between 30 and 10 arc-minutes, more than enough to cover the Sun and part of its corona. The apparent diameter of Callisto, however, is practically equal to that of the Sun; its eclipses are thus quite comparable with a total eclipse of the Sun as seen from Earth.

Io completes an orbit around Jupiter in less than two days, Europa in slightly more than three days, Ganymede a week, and Callisto in just two weeks. From this we can see that eclipses – always total – are very frequent on Jupiter! It is always a fascinating sight for amateur astronomers, who, on some nights, may be able to see one, two or even three total eclipses occurring on the surface of another world. More rarely, two or three eclipses occur at the same time! The sight of an eclipse on Jupiter, in a good tele-scope, is striking: the immense globe of the planet, banded with white, yellow and ochre-coloured clouds, is slowly crossed by a tiny black shadow. In most cases, the satellite responsible for the eclipse is also visible, like a tiny point of light. With Jupiter, of course, every eclipse is followed by an occultation! Once a satellite has passed in front of the changing surface of the giant planet, the latter, in turn, occults the satellite. Finally, each satellite's orbit ends with an eclipse, when the satellite, after having been occulted, re-emerges from Jupiter's shadow. When the Earth lies outside the Sun–Jupiter line, the effects of perspective cause us to see the satellite emerge gradually from the shadow, at a consider-able distance from the planet's brilliant disk.

Knowing the orbital periods of the Galilean satellites, ever since the beginning of the 17th century, astronomers have had no difficulty in predicting the date and time of eclipses. In 1672, however, systematically observing these eclipses from the Paris Observatory, the young Danish astronomer Ole Rømer noted a strange discrepancy between the theoretical predictions and his own observations. When the Earth was most distant from Jupiter – i.e., on the opposite side of the Sun from Jupiter – the eclipses showed a delay of eight minutes relative to the predic-tions. When, on the other hand, the Earth and Jupiter were on the same side of the Sun, and thus at their closest, the eclipses occurred eight minutes early! Ole Rømer quickly recognized the origin of this periodic and systematic difference. He was the first to realise that light does not travel at an infinite velocity. What he was observing was simply the difference in the time

■ That day , 23 April 1998, on Ascension Island, the Moon first uncovered Venus (top) and then Jupiter. The two bodies are clearly visible, but look as if they are squashed right up against the disk of the Moon...

■ AT DAWN, ALL IS OVER: BETWEEN THE CLOUDS, THE MOON AND THE TWO PLANETS SEEM TO SMILE AT THE FEW SPECTATORS OF THIS ASTRONOMICAL SPECTACLE. WE SHALL NOT SEE A SIMILAR PHENOMENON FOR ANOTHER CENTURY OR TWO.

light took to travel the distance from Jupiter, alternately with, and without, half the diameter of the Earth's orbit, that is, about 150 million kilometres. Rømer was thus the first to measure the velocity of light: about 300 000 km/s.

Just like the Moon, the planets appear to follow a great circle around the celestial sphere. For some of them, of course, a complete circuit takes a long time: more than 11 years for Jupiter, 84 years for Uranus, and nearly 165 years for Neptune. In their long solitary journey, again like the Moon, the planets occasionally encounter a star. If the latter is sufficiently bright not to be overwhelmed by the brilliance of the planet, a stellar occultation can be extremely interesting for astronomers. This is because, when a giant planet with a dense atmosphere passes in front of a star, scientists are able to study the extinction of the star by the gaseous limb of the planet. Calculation then enables them to determine the density and even the temperature of the atmosphere through which the starlight momentarily passes. And then, if they have sufficiently powerful telescopes, researchers are able to go even farther. Spectral analysis of the light from the star, possibly carried out as the latter slowly fades, provides valuable information about the chemical composition of the atmosphere: specifically, absorption lines from the planet's gases are superimposed on the stellar spectrum. This is why, for several decades, astronomers have tried not to miss the occultation of a star by a planet.

On 10 March 1977, various teams therefore prepared to follow a ninth-magnitude star in the constellation of Libra, SAO 158687, as it was occulted by the planet Uranus. It was an event that would be particularly favourably visible from the Indian Ocean, which was why the team led by James Elliott were on board NASA's flying observatory, the Kuiper Airborne Observatory. Flying at an altitude of 12 000 m, a major surprise awaited the astronomers. Forty minutes before the occultation, the astronomers saw SAO 158687 suddenly disappear for a few seconds, reappear, and then vanish and reappear eight more times in succession at regular intervals, before disappearing behind the disk of the planet. After the occultation, the same phenomena occurred in the opposite order: the nine

■ The satellite Io, although similar in size to the Moon, appears minute when it transits the disk of Jupiter, as observed here by the Hubble Space Telescope. The satellite's shadow-cone is projected onto the giant planet's clouds, creating a spectacular total eclipse.

rings around Uranus had been discovered. Too thin, too dark, and too narrow, they had previously escaped detection, but the transient eclipse of SAO 158687 that they produced enabled the determination of their diameters, their widths, and their density.

This fine discovery soon gave astronomers the desire to try their luck with the planet Neptune. The last of the giant planets was thus observed systematically every time it occulted a sufficiently bright star. At the beginning of the 1980s, no one understood what was happening during these occultations: either nothing at all was seen, or else there was a brief occultation before the occultation by the planet, and nothing afterwards. Nothing at all seemed to resemble a system of rings. But then, thanks to the occultations, scientists finally understood what they were observing; Neptune was indeed, like Jupiter, Saturn and Uranus, surrounded by a system of rings, but these, apart from being extraordinarily narrow and thin, were, in addition, irregular: only the densest portions were capable of producing a stellar occultation!

There are other eclipses, which are extremely numerous, very inconspicuous, and invisible to the naked eye, unknown by the general public and poorly known to most amateur and professional astronomers. To specialists, they are, nevertheless, a perpetual source of discoveries. These are occultations of stars by asteroids. These events are visible only with telescopes of respectable size and, are, in fact, only partly predictable. Although the paths of thousands of asteroids against the sky are known extremely accurately, because the size or true diameter of these minute planetary fragments are unknown, it is impossible to predict whether they will, or will not, occult a star that happens to lie on their path. As a result, ingenious astronomers worked out how to use these unlikely eclipses precisely for determining the diameter and the shape of the asteroids. This work is often undertaken by groups of amateurs, who, to succeed, need nothing more than a telescope, good eyesight, a tape recorder, and a stopwatch. When a possible stellar occultation by an asteroid is announced for some area of the Earth, these groups try to observe it, from several sites, several tens of kilometres from one another. If one particular observer is too far north and another too far south, they will see nothing at all. Observations obtained between these two extremes, on the other hand, after calculation allow one dimension of the asteroid (which is generally irregular) to be determined precisely. Similarly, recording the duration of the occultation from several sites gives another of the object's dimensions. This technique has been remarkably successful, as shown, for example, by precise measurements of the asteroids 105 Artemis and 39 Laetitia, made during occultations on 4 December 1997 and 21 March 1998, respectively. Artemis was found to be an irregular body 98 km by 118 km, and Laetitia a respectably sized, elliptical object of 219 km by 143 km. Such accurate measurements are not possible even with giant telescopes: to them, asteroids are too distant and too small to appear as more than simple points of light.

■ THE COMPLEX INTERPLAY OF SHADOWS CAUSED BY THE TRANSIT OF SATELLITES IN FRONT OF JUPITER, OBSERVED OVER SEVERAL HOURS FROM AN OBSERVATORY IN NEW MEXICO. *BOTTOM LEFT*: THREE TOTAL ECLIPSES ARE OCCURRING SIMULTANEOUSLY ON THE GIANT PLANET!

TRANSITS OF THE SUN

Although planets and asteroids are perfectly capable of occulting distant stars, it is quite impossible for them to hide the closest star: the Sun. This very specific form of eclipse does, however, occur relatively frequently in one or other region of the Solar System. We, on Earth, are not particularly spoilt in this respect, because, our planet being the third from the Sun, we can see only Mercury and Venus pass in front of it. These events – known as transits to astronomers – are rare eclipses, invisible to the naked eye. During a transit, Mercury and Venus appear as just a small black point (12 and 62 arc-seconds across, respectively) against the blinding disk of the Sun, which should be observed only with a telescope fitted with a properly designed, dense filter. Since the invention of the telescope by Galileo at the beginning of the 17th century, just 55 transits of Mercury have taken place, the last being on 15 November 1999. As seen from Earth, Mercury has an apparent diameter that is about one two-hundredth of that of the Sun. Transits of Venus, much rarer, are also much more spectacular – as seen from Earth, the planet measures about one thirtieth of a solar diameter – and are awaited like some small astronomical miracle by eclipse chasers. In fact, most of them, during the course of their lifetime, are never lucky enough to witness one of the events. Since the time of Galileo, only six transits of Venus have occurred, in 1631, 1639, 1761, 1769, 1874, and 1882. Since then – nothing. There was not a single transit of Venus in the 20th century. The next transits will occur on 8 June 2004, 6 June 2012, 11 December 2117, and 8 December 2125.

In the history of astronomy,

■ Every 15 years, a series of total eclipses of the Sun occur on Saturn. Here, beneath the Hubble Space Telescope's eagle eye, on 6th August 1996,

transits of Venus have had considerable significance. The first of them, that of 24 November 1639, was followed from England by Jeremiah Horrocks, who had predicted the event by computation. For astronomers, the following transit – for which they had to wait 122 years! – was of crucial importance. Edmond Halley had, in fact, devised a method of calculating the size of the Solar System, unknown at that time, by using the transits of Venus. It involved astronomers observing the event from several sites as distant as possible from one another. Venus would pass in front of the Sun with a different, very slight shift for each observer. Measuring this angle against the Sun – known to specialists as the solar parallax – should, knowing the distance between the observers, enable the exact Earth–Sun distance (known as the 'astronomical unit') to be calculated. From that, and Kepler's laws, astronomers would finally be able to determine the true scale of the Solar System.

As one might imagine, the transits of 1761 and 1769 were impatiently awaited. Dozens of different teams, with their telescopes and clocks, were sent to various parts of the world. The measurements obtained by these expeditions did enable the value of the astronomical unit to be refined, but were sometimes carried out under dramatic circumstances. In 1768, Jean d'Auteroche and his astronomer colleagues crossed the Atlantic and Mexico, finally successfully observing the transit of 6 June 1769 from Baja California. Remaining a few days more to study a lunar eclipse, the group was struck by illness. D'Auteroche observed the eclipse as he was on the verge of death, and three other astronomers also died. Only the engineer Jean Pauly remained to bring the precious measurements back to Paris. Then again, there was the astronomer Jean-Baptiste Le Gentil de La Galaisière, who, in March 1760, embarked on a ship of the East India Company, to observe the transit of Venus that was to occur fifteen months later at Pondicherry. But, having arrived just a few cables' lengths from the French trading post, he was unable to disembark because of the war that was raging between the French and the English, and on

■ The rings of Saturn's are plunged into the shadow of the planet. Above them is the satellite Tethys. Another satellite, invisible here, casts its shadow onto the planet.

Titan, a satellite larger than the Moon, casts its shadow on the ringed planet.

6 June 1761, he had to observe the transit from the deck of the ship. His clocks could not be used at sea, so his mission was a failure. But Jean-Baptiste Le Gentil de La Galaisière was a forward-looking man. He decided to remain in the East Indies until the following transit of Venus, which was to take place in the region 'only' eight years later, on 3 June 1769. Returning to Pondicherry, where the war had finished, Le Gentil settled down and waited. On 3 June, finally ready after years of waiting, with all his instruments trained on the Sun, he saw clouds spread across its face Sun just as the transit took place. He returned to France in 1771, ill, exhausted, after twelve years' absence. But his ordeal was not over. On arrival at the Académie des Sciences, he found that his post had been given to another astronomer. For their part, his heirs, having declared him dead, had divided all his goods between them! Le Gentil's story did, however, have a happier ending; he recovered from his tropical fever, then married. Finally, his post as astronomer was restored by the Académie, which published his *Voyage to the East Indies (1760–1771) Undertaken on the Orders of the King, on the Occasion of the Transits of Venus across the Disk of the Sun, 6 June 1761, and 3 of the same Month 1769.*

The two following transits, those of 1874 and 1882, allowed astronomers – there were several hundred who were passionately determined to observe the event! – to confirm what they had suspected for a long time: the planet Venus is surrounded by a thick atmosphere. In fact, at the moment when the planet just touched the limb of the Sun, its disk was surrounded by a diffuse luminous ring. This was undoubtedly caused by sunlight, refracted by the upper atmosphere of Venus.

For any planet in the Solar System, the transit of a planet with a smaller orbit is visible. One day, perhaps, astronauts on Mars will see the Earth and Moon transit the disk of the Sun. Although the conquest of Mars, promised for over twenty years by the various space agencies, has regularly been put back from one

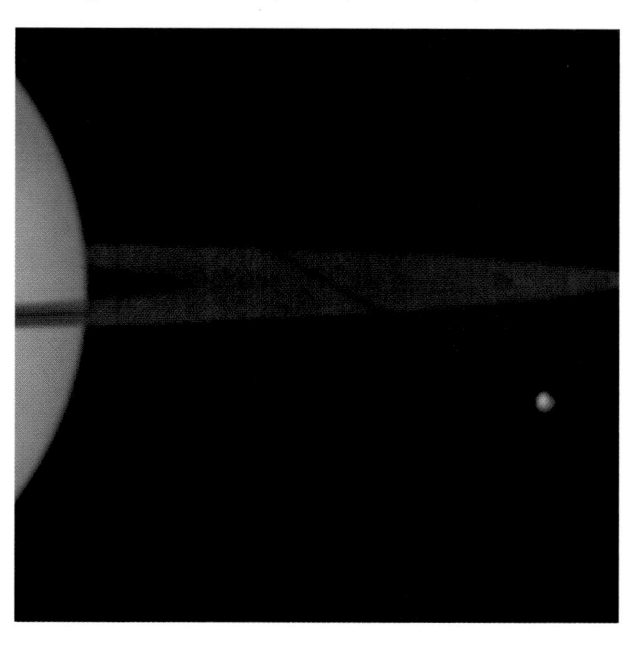

■ This narrow line running across Saturn's rings is the shadow cast by the satellite Dione. An astonishing eclipse, accidentally caught by the Hubble Space Telescope.

■ On 2 July 1989, Saturn passed slowly in front of the star 28 Sagittarii. The star was occulted by the icy rings, but appears here for a brief moment, shining through the practically empty region known as the Cassini Division.

decade to another, we can still dream that someone will observe the next transits of the Earth–Moon pair, predicted to occur on 10 November 2084, 15 November 2163, 10 May 2189, 13 May 2268, and 13 November 2368...

See you in seventeen centuries

But will someone, some day, see a transit of Jupiter from Saturn or one of its numerous satellites? According to the calculations carried out by the Belgian Jean Meeus, one of the current experts in celestial mechanics, just one is predicted to occur in the next two millennia: 29 October 3728. Finally, we will not have a second chance to observe the sight, like something out of science fiction, offered by the transit of Saturn, as seen from Uranus. This exceptionally rare event, which has not occurred in the past two millennia, will take place just once in the next two, on 8 April 2669.

As we have seen, all the planets in the Solar System move in approximately the same plane, the vestiges of the disk of gas and dust from which they condensed around the Sun, 4.5 billion years ago. As seen from the Earth, this plane appears like a broad ribbon encircling the celestial sphere. This is the Zodiac, which nominally consists of twelve constellations. It is here that the Sun, the Moon, the nine planets and their sixty-odd satellites eternally revolve in their heavenly clockwork, projecting their long shadows into space, which occasionally brush across one body or another, plunging it into darkness for an instant.

But eclipses are not confined to the Solar System. The whole cosmos is like a vast shadow play, where even stars occasionally eclipse one another. In the constellation of Perseus, there is a bright star, called Algol – the demon star – by Arab astronomers. Approximately every three days, the magnitude of this star drops in a spectacular manner, before returning to its normal brightness. It was John Goodricke, in 1782, who was the first to suggest that Algol was regularly subject to eclipses by another star orbiting around it. Subsequently, this theory was confirmed. Algol has been minutely observed. The unusual characteristics of the pair of stars involved and the eclipses that occur, are now perfectly well understood. Algol is a double star, some 95 light-years from the Earth. The primary component is a giant star, 4 million kilometres in diameter, and one hundred times the brightness of the Sun. The less massive companion measures nearly 5 million kilometres in diameter, and its brightness is about twice that of the Sun. Algol's primary is about five times the Sun's mass, and the secondary has about the same mass as the Sun. These two giant stars, separated by just 10 million kilometres, are orbiting at a ridiculous rate: it takes exactly 2 days, 20 hours, 48 minutes and 56 seconds for the companion to complete one orbit of the main component! Because the orbital plane of Algol is only slightly inclined to our line of sight, every orbit we see the fainter star eclipse the brighter. The result is the dramatic drop in Algol's luminosity, which, for some tens of minutes, appears about one third of its normal brightness. Astronomers nowadays have catalogued thousands of stars like Algol in our Galaxy, and which are known as eclipsing binaries. Many are kept under constant surveillance, in order to understand, amongst other aspects, the complex gravitational interactions linking these extremely close binaries, where both bodies are constantly evolving.

There remains the most extraordinary, the most mythical, and yet the most important eclipses, which no one has ever seen, and which astronomers dream of seeing one day. For some years now, they have been devoting more and more resources to this end. In our Galaxy there are, at the most conservative estimate, two hundred billion stars. We now know that 5 per cent of them, at least, are surrounded by a system of planets. By the end of the 20th century, more than twenty of these extrasolar planets had been discovered. With the progress in astronomical instrumentation, astronomers have come up with a new method of searching for these planets: they want to observe distant eclipses of stars caused by the planets orbiting them.

■ A transit of Mercury in front of the Sun, observed on 6 November 1993 from Australia. Although not spectacular, but very rare, these transits, like those of Venus, are impatiently awaited by astronomers.

Eclipses among the stars

For several years an international monitoring network has been searching for these transits by extrasolar planets, using moderately sized telescopes, equipped with extremely sensitive electronic detectors. So far, no convincing result has been obtained. Two stars, CM Draconis and Beta Pictoris, may have been the site of a transit, but it is now impossible to verify this. One of the difficulties for astronomers, is, of course, how to distinguish a possible transit – which will appear as only a very slight dimming of the star being monitored – from any intrinsic variation in the brightness of the star itself, caused, for example, by a drop in the temperature of its photosphere. Then again, they need to rule out possible artefacts, or false alarms, caused by unstable optics or electronics, or by unforeseen variations in the Earth's atmosphere. Finally, for this method of research to succeed, they need, above all, to monitor a large number of stars. As we have seen for the Solar System, planetary transits are rare: in the case of a terrestrial-type extrasolar planet, it would take place over just six hours – once a year! Then again, statistically, the probability that an extrasolar planet will eclipse its star is just 1 per cent. Indeed, in most cases, planetary systems around other stars will not cause eclipses, because the orbital planes do not lie in our line of sight. In the light of all these difficulties, various scientists, including the French astronomer Jean Schneider, have suggested to the various space agencies that the hunt for extrasolar planets should be conducted from space. Outside the atmosphere, stars do not scintillate, so their brightness is perfectly steady, their possible intrinsic variations are easier to identify, and, as a result, the observation of extrasolar eclipses should be much simpler. This is why, in 2002, CNES, the French space agency is to launch an astronomical satellite, Corot, designed to study, on the one hand, the ways in which stellar envelopes vibrate and, on the other, to monitor continuously fifty thousand stars, in the hope of finally witnessing an extrasolar eclipse. Placed into a polar orbit with a period of 150 days, Corot, will not take its eyes off the stars. If an extrasolar planet passes in front of its star, the brightness of the latter will diminish, depending on the size of the planet, by 1/1000 to 1/10 000, and last for several hours.

Naturally, to confirm the discovery of an extrasolar planet,

■ The play of light and shade in eclipses is universal. Stars eclipse one another, and planets eclipse stars around which they orbit. In the decades to come, astronomers will be monitoring tens of

we shall have to wait until a second eclipse occurs, after the planet has completed one orbit of its star. In practice, then, it is extrasolar planets that are close to their star, like Mercury, Venus, and the Earth in our system, which are most likely to be discovered by this method. Astronomers express the hope of finding 50 to 100 extrasolar planets, more or less resembling the Earth, during the Corot mission. Will one of those extrasolar transits perhaps be double? Caused by a planet and its satellite, gears in some other cosmic clockwork, that will, in turn, create from time to time, strange and marvellous eclipses on other worlds, unseen by any eyes... ■

THOUSANDS OF STARS, IN THE HOPE OF DISCOVERING NEW WORLDS, DETECTED BY THE ECLIPSES THAT THEY CAUSE TO THEIR STAR...

By the light of eclipses

■ DURING A TOTAL ECLIPSE, THE SUN'S INVISIBLE EMPIRE APPEARS: ERUPTIONS AND PROMINENCES RISE ABOVE THE LEVEL OF THE PHOTOSPHERE, AND COLOUR THE SUN'S INNER ATMOSPHERE, KNOWN AS THE CHROMOSPHERE. THIS IMAGE WAS OBTAINED AT THE HALEAKALA OBSERVATORY, ON HAWAII.

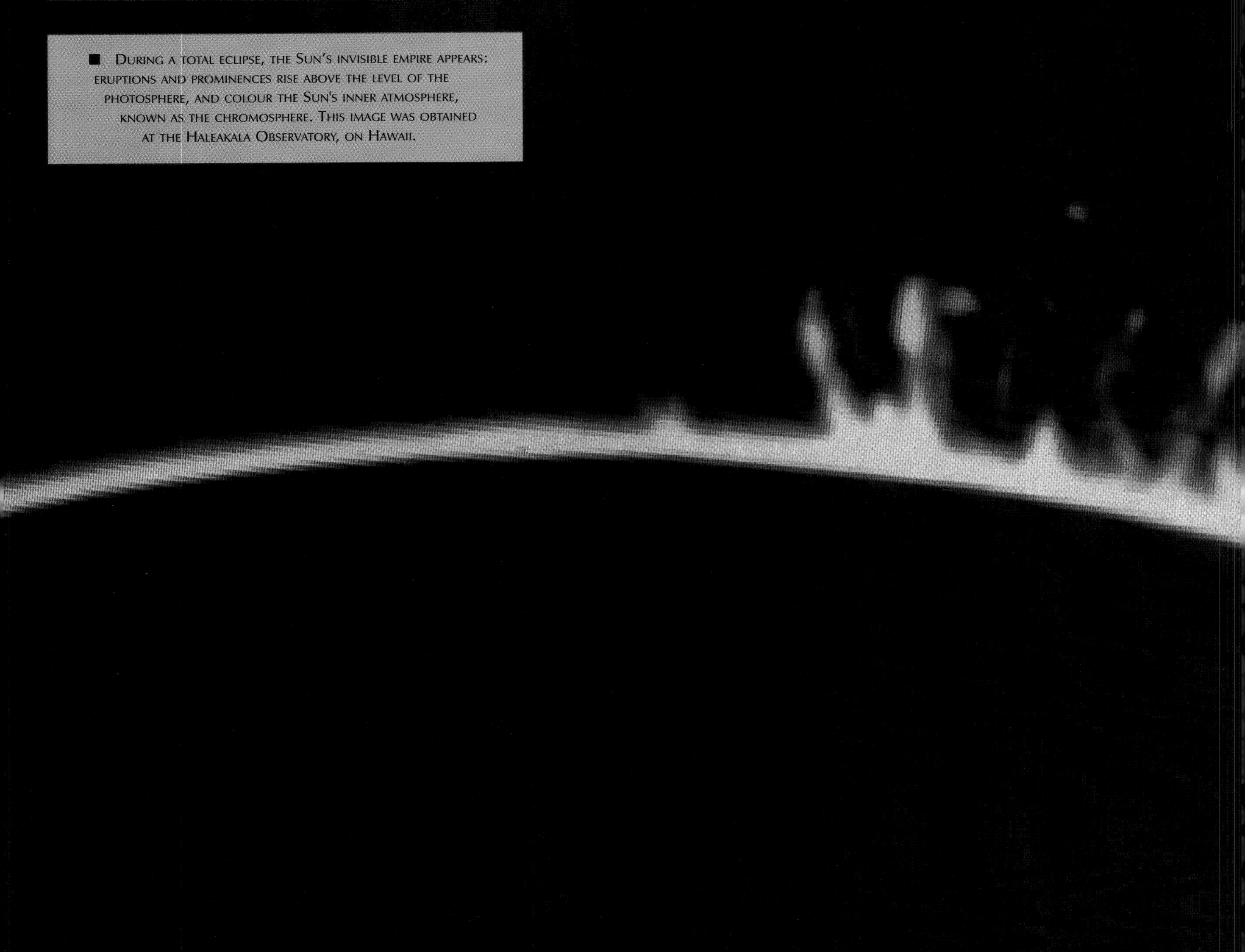

■ During the course of this total solar eclipse, observed from Mexico, a loop-shaped prominence rose from the chromosphere, while the Sun's outer atmosphere, the corona, appeared as a more or less regular halo.

Eclipses of the Sun and Moon have never ceased to provide us with a host of lessons about the nature of the universe around us. The first of these lessons concerned the celestial bodies directly involved in eclipses: namely the Earth, Moon, and Sun. Indeed, back in antiquity, the proof that the Earth was round, and the first measurements of the respective sizes and distances of the Moon and Sun were deduced from the observation of eclipses. In the 19th century, it was the normally invisible atmosphere of the Sun that was revealed thanks to eclipses. Far from being the perfectly round, and sharply defined ball of hot gas that it appears to the eye – appropriately protected by suitable filters, of course – the Sun is found to be a sprawling giant, overflowing with energy, plasma, and particles, that extends its influence throughout the whole Solar System. Eclipses also provoked the discovery of helium, the second most abundant element in the Sun, and in the universe as a whole. In a more surprising manner, in the 20th century, Einstein's General Relativity, a fundamental theory about space, was tested experimentally for the first time, thanks to an eclipse. It is on this new vision of the universe, which explains gravitation in terms of the 'curvature of space-time', that all our current knowledge of the origin, the structure, and the evolution of the universe, depends, by way of the fascinating concepts of an expanding universe, the Big Bang, and black holes.

The Earth is round

The first demonstration of an astrophysical nature resulting from eclipses is the one given by Aristotle concerning the fact that the Earth is round. The astronomical views of this Greek philosopher are well-known to us, thanks to his two works, known to us as *Meteorology* and *On the Heavens*, dating from the 4th century BC. Like other thinkers of his day, Aristotle believed that all heavenly bodies were spherical, because to him heavenly bodies were a reflection of divine perfection, and the sphere is the most outstandingly perfect geometrical figure. But this argument was not a physical demonstration, because, naturally, Aristotle did not have any experimental means of confirming the spherical nature of the planets and stars.

As far as the Moon was concerned, the philosopher adopted an explanation attributed to the Pythagoreans, namely that the observed appearance of the Moon throughout its various phases corresponded to a spherical body, half of which is illuminated by the Sun. As for the spherical nature of the Earth, the proof given by Aristotle is quite original: he notes that an eclipse of the Moon is caused by the shadow of the Earth, and that the circular

■ The total eclipse of the Sun of 26 February 1998. The electrified plasma hugs the lines of the Sun's magnetic field, just as iron filings follow those of a magnet. The corona displays narrow filaments over the poles, and extends in a more homogeneous manner in the equatorial regions.

shape to the edge of the shadow seen on the Moon's surface implies that our world is spherical.

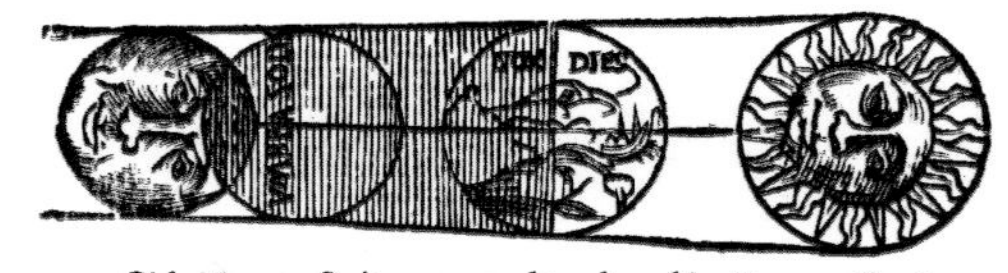

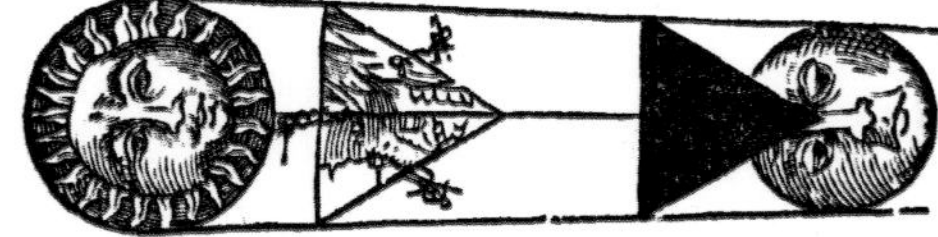

■ According to Aristotle, lunar eclipses prove that the Earth is round. Indeed, if the Earth were square, or triangular, its shadow projected onto the disk of the Moon at the time of an eclipse would not appear circular.

Sizes and distances of the Moon and Sun

The golden age of Greek astronomy flourished at Alexandria. Since its foundation under the reign of Ptolemy Soter (3rd century BC), the Alexandrian school brought together brilliant mathematicians and geometers, such as Euclid, Archimedes, and Apollonius. Similarly, the greatest ancient astronomers Aristachus of Samos, Eratosthenes, and Hipparchus, as well as Ptolemy (2nd century BC), all worked there.

Aristarchus (310-230 BC) is nowadays known for having been the first to voice the heliocentric theory, i.e., that it is the Sun that reigns at the centre of the world system, not the Earth as was believed at the time. His statement does not appear in any known work, but it was reported by Archimedes and by Plutarch. The only work of Aristarchus that has come down to us relates to the sizes and distances of the Sun and the Moon.

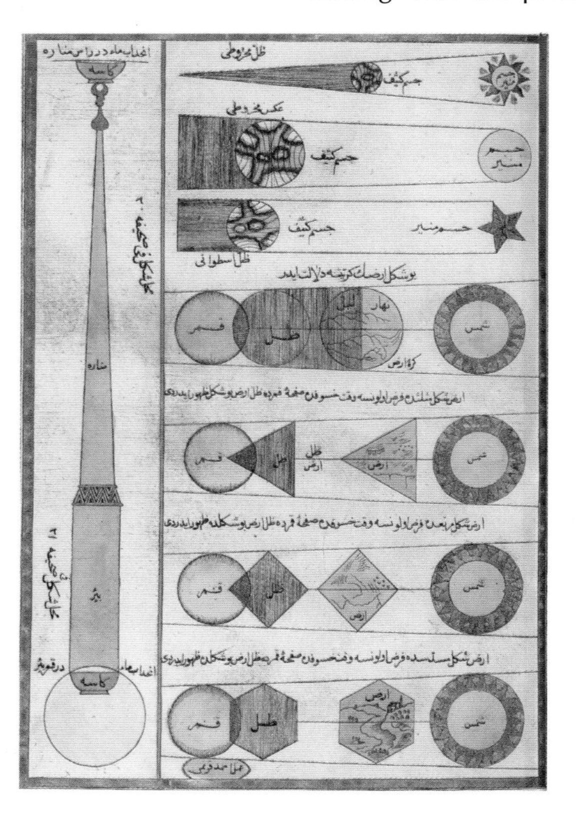

■ Aristotle's geometrical argument is shown in several ancient astronomy texts, including *Cosmographia*, by Petrus Apianus and Gemma Frisius, 1581 (*top*) and Mustapha Ibn Abdullah's *Book of the Description of the World*, 1732 (*bottom*).

The Alexandrian astronomer completely reopened this question, which had been discussed since the 4th century BC. The Pythagoreans had positioned the heights of the celestial bodies according to musical intervals. Eudoxus, the brilliant disciple of Plato, had estimated the diameter of the Sun as nine times that of the Moon. As for Aristarchus, he devised an ingenious geometrical method of calculating the distance ratios of the Sun and Moon. He found that the Sun lay at a distance between 18 and 20 times that of the Moon. (In fact, it is 400 times as far.) By an argument based on the observation of eclipses, he determined the diameter of the Moon as one third of that of the Earth, which is very close to the actual value. He also announced that the diameter of the Sun is seven times that of the Earth. Even though Aristarchus considerably underestimated the size of the Sun, because it is actually 109 times as large as the Earth, he had grasped the essential fact that the daytime star was much larger than the Earth. It was precisely this result that led him to the heliocentric hypothesis. He did, in fact, argue that under these circumstances, it was logical to believe that the Earth and the other celestial bodies revolved around the Sun, rather than the reverse.

Aristachus was before his time. The world had to wait until 1543 and the work by Copernicus, before the heliocentric theory was again put forward, this time with success.

A century after Aristachus, and again at Alexandria, Hipparchus developed a complete theory of the Moon. He defined the lengths of the synodic month (or lunation, the period in which the Moon returns to the same position relative to the Sun); the draconitic month (the period for the Moon to return to the same position relative to the nodes of its orbit); and the anomalistic month (the period for the Moon to return to perigee or apogee). The immense improvements that Hipparchus brought to theories of the apparent motion of the Moon and Sun enabled him to have far more success than his predecessors in dealing with the problem of predicting eclipses, which had always been of immense interest. Hipparchus considerably extended Aristarchus' method: by observing the angular diameter of the shadow of the Earth at the Moon's distance during a lunar eclipse, and comparing it with the known apparent diameters of the Sun and Moon (about half a degree), he obtained the ratio of the Earth-Moon and Earth-Sun distances, giving one when the other is known. Pappus, another famous astronomer of the Alexandrian school, recounts that Hipparchus made the following observation of: 'an eclipse of the Sun, which in the area of the Hellespont was precisely an exact eclipse of the whole Sun; such that none of it was visible, but at Alexandria, in Egypt, about 4/5 of its diameter were hidden. By means of the foregoing arguments, [Hipparchus] showed that, measured in units where the radius of the Earth has the value of 1, the smallest distance to the Moon is 71, and the largest 83. Whence the average of 77.'

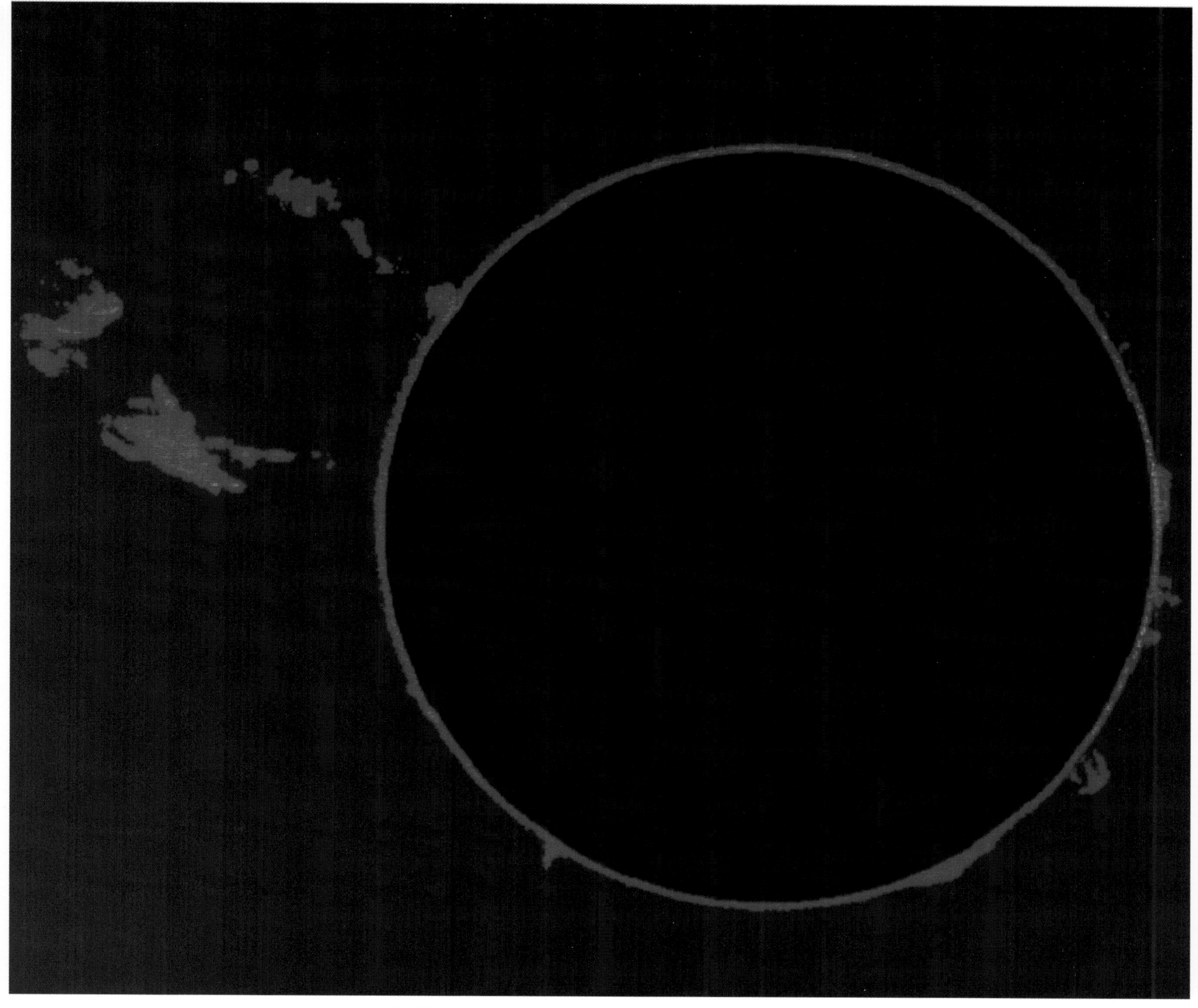

■ Major solar eruptions, visible during solar eclipses, may hurl plasma a million kilometres above the chromosphere.

The total solar eclipse mentioned is that of 20 November 129 BC. The actual value of the Earth-Moon distance is 60.4 terrestrial radii.

The empire of the Sun

The name 'photosphere' is given to the layer of gas that provides most of the light from the Sun. Its limit, seen in profile, forms the apparent edge of the disk. During a total solar eclipse, however, the Sun's invisible empire is revealed: its two outer atmospheres, the chromosphere and the corona, appear as a irregular halo.

The chromosphere extends from the photosphere up to a height of about 15 000 km. Its name arises from the fact that the emission lines in its spectrum were originally detected in different 'colours' of the visible spectrum, which – as we may recall – may be dispersed to give the colours of the rainbow, from red to indigo. The chromospheric lines are red, green, blue, and violet from hydrogen, and yellow from helium. The chromosphere is the region in which eruptions and prominences occur.

As regards the corona, it is the outer, highly tenuous, region of the Sun's atmosphere. It may be sub-divided into two layers, the inner corona between 15 000 km and 200 000 km above the photosphere, and the outer corona beyond that. It consists of negatively charged electrons, and positively charged atoms known as ions. Under normal conditions, atoms are electrically neutral, but when they are raised to high temperatures, they lose their electrical neutrality together with their electrons, and are ionized. This is what happens in the solar corona. This flux of charged particles is ejected into interplanetary space by the Sun, and forms the solar wind. When it reaches the vicinity of the Earth, the solar wind is still blowing at about 400 km/s. Its particles are channelled by the Earth's magnetic field and may cause perturbations – the notorious geomagnetic 'storms' – and polar aurorae.

The corona consists of plumes, filaments and streamers of pearly white light. Although it is one of the most beautiful and spectacular aspects of the Sun, it is so pale that it is invisible under normal conditions. In fact, although the corona is extremely hot – two million degrees – it is also very tenuous. In visible light it radiates about one millionth of the light from the main surface, the photosphere. It is, as a result, completely drowned by the latter's brilliance, so that none of it whatsoever may be seen.

A total eclipse of the Sun is an astronomical event that offers the possibility of seeing the corona in all its splendour. At the precise moment when the disk of the Sun vanishes, this incandescent halo suddenly appears, like a vast fan of light, with a diameter about twice that of the solar disk. Some of its jets may be followed with the naked eye out to around six solar radii. This is the reason why total eclipses are so sought-after by professional solar physicists, who do not hesitate to organise expeditions to the other side of the world to take advantage of those few moments of observation.

The true nature of the solar atmosphere was not understood until the second half of the 19th century. Before then, the existence of a luminous halo visible only at eclipses was mentioned by Plutarch in the year 98, and discussed by Kepler in his *Epitome* on Copernican astronomy in 1621. A coloured halo was also described by Halley during the eclipse of 1715, in the form of a narrow red band. But astronomers all made the error of thinking that it was either an optical illusion, scattering caused by the Earth's atmosphere, or evidence of a lunar atmosphere.

The idea of an atmosphere of the Sun itself took shape in the 18th century, with Dortous de Mairan and Cassini. Their observations were not of eclipses but concerned the zodiacal light. The zodiacal light is a faint, night-time glow, which may be detected on a clear, Moonless night. Centred on the plane of the ecliptic – the Sun's apparent path across the celestial sphere, during which it crosses the constellations of the Zodiac, whence the term – the zodiacal light is nowadays understood to be caused by the scattering of sunlight by the cloud of interplanetary dust that surrounds the Sun. In this way some of the regions of the Sun's empire that are ordinarily hidden come into view.

The first exhaustive study devoted to observation of the solar halo was carried out by François Arago, at the total eclipse of 1842. Clearly describing the presence of the halo, he suggests that it should be called the 'lunar corona', thus perpetuating the error that had already been committed by his distinguished predecessors.

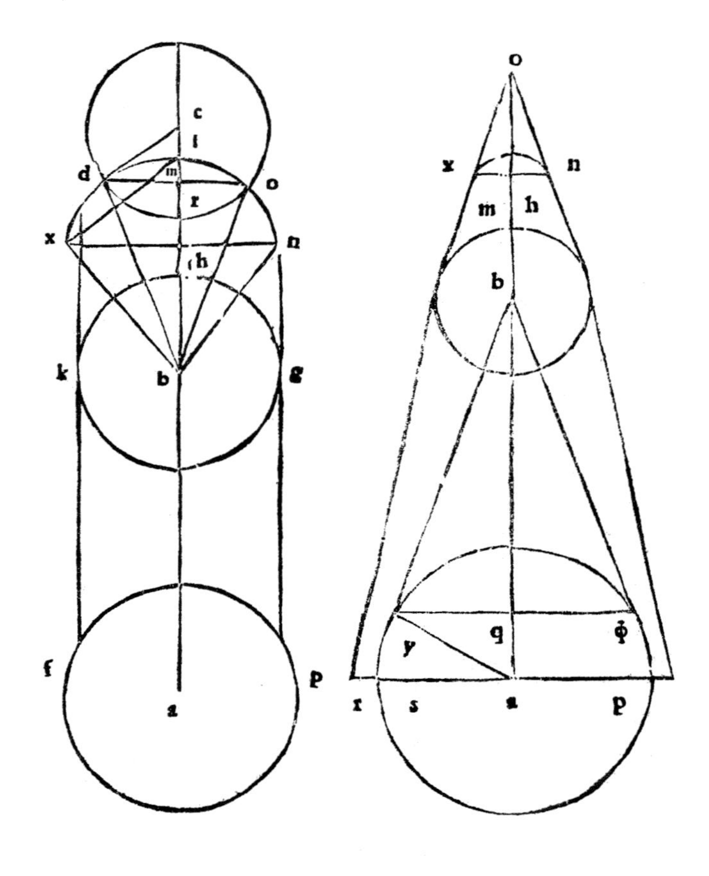

■ Aristarchos of Samos tried to calculate the relative diameters of the Moon and Sun, as deduced from the line subtending the arc that divides the light and dark portions of the Moon during an eclipse.

It was not until the eclipse of 28 July 1851 that the halo, photographed for the first time by Berkowski, was definitely shown to be physically associated with the Sun. The situation was confirmed in 1860, during an eclipse observed from Spain. Warren de la Rue and Angelo Secchi obtained daguerrotypes of the solar halo from two sites on the path of totality, but 500 km apart. The structures seen during the eclipse were similar on both photographs. There was, therefore, no parallax effect. Parallax is the change in perspective when an object is seen from two different positions. If you hold your finger up in front of your nose and alternately close the right and left eye, your finger moves relative to the background. If you increase the distance to your finger, by extending your arm, and try it again, the finger still moves, but less distinctly: the farther the finger is away, the smaller the movement. The experiment by de la Rue and Secchi observed solar prominences with eyes 500 km apart. The absence of parallax proved that these structures did not belong to the Moon, which is too close, but to the Sun, which was sufficiently far away for the change in perspective to be imperceptible.

During the total solar eclipse of 18 August 1868, the spectrum of the prominences and chromosphere was obtained for the first time by Jules Janssen, the future founder of Meudon Observatory, and by Norman Lockyer. The spectrum proved to consist of bright lines, demonstrating that the source of the emission consisted of gas. The most important lines present in the spectrum of the chromosphere were those of hydrogen, two lines belonging to calcium, and a yellow line of a hitherto unknown element. Believing that it was an element unique to the Sun, it was called 'helium' (from the Greek *helios*, Sun). Helium was discovered on Earth in 1895. Air contains 0.0005 % by volume. It is a 'noble gas', very light, and therefore used to inflate balloons. Although it is rare on Earth, helium is, in fact, the second element in the periodic table and is, after hydrogen, the most abundant element in stars and the universe as a whole: it amounts to 25 % of the total mass.

During the course of a solar eclipse, at the exact moment when the western edge of the

fig 1. Protubérances le 19 Août 1868.
le jour de la découverte de la méthode
Observations faites à Guntoor, Inde.
puis Masulipatam.

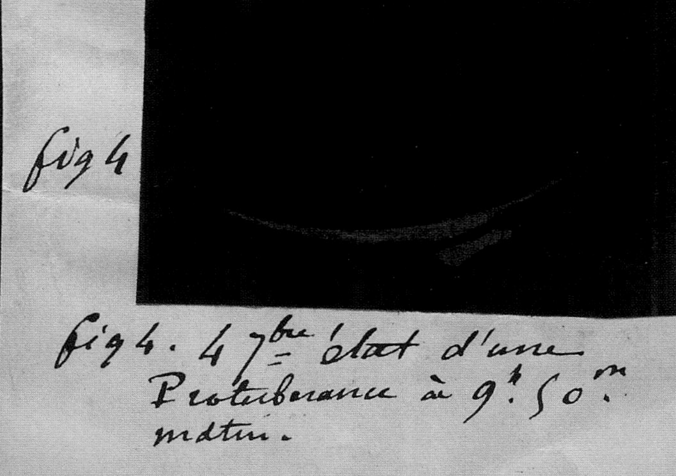

fig 4. 4 7bre état d'une Protubérance à 9h 50m matin.

fig 5. même Protubérance à 10h 50m (1h après).

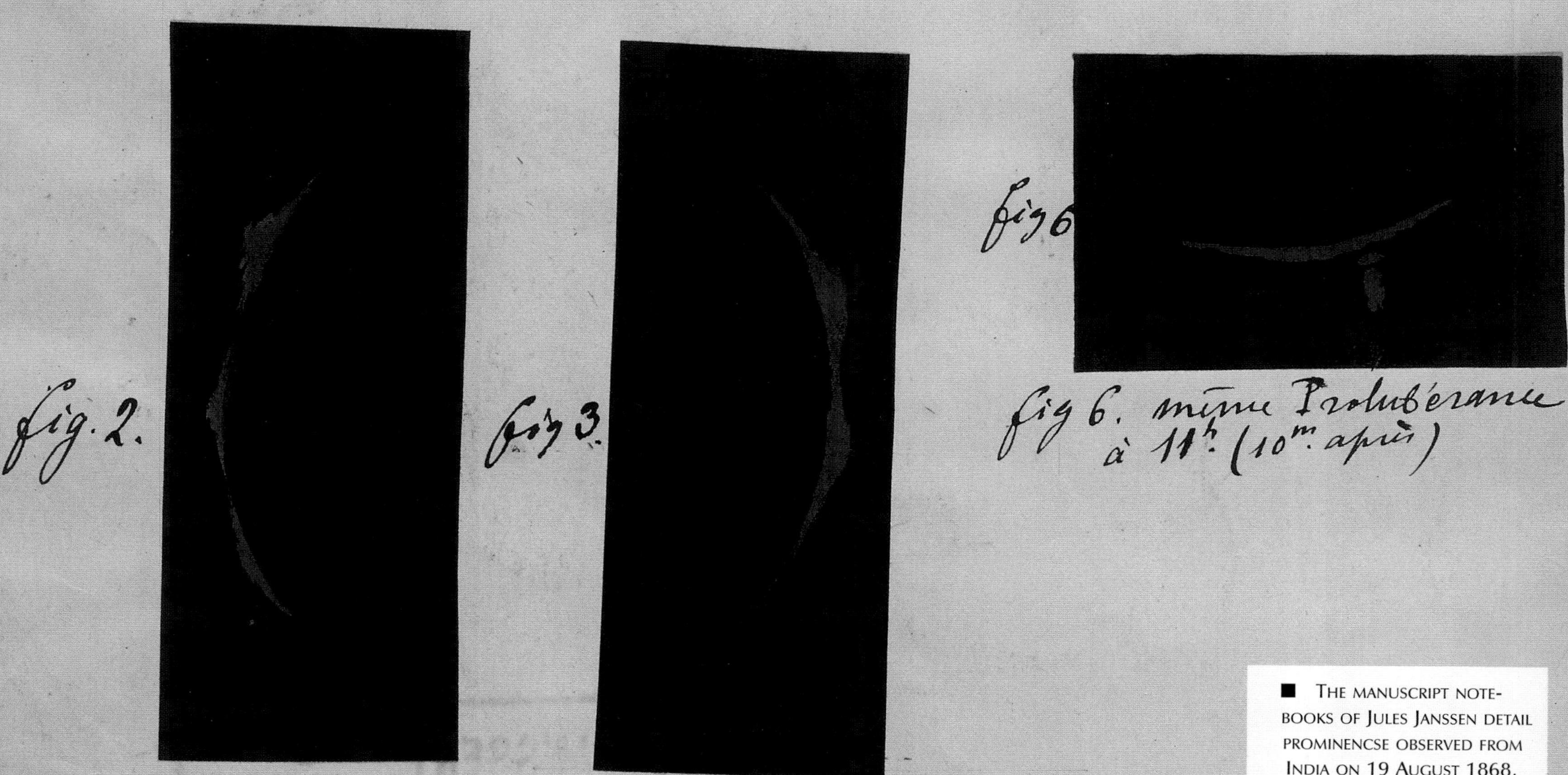

fig 6. même Protubérance à 11h (10m après)

fig. 2. Protubérances sur le bord oriental du Soleil, le 20 Août 4 heures du soir.

fig 3. Protubérances sur le bord occidental du Soleil le 23 Août, 10h 1/2.

■ The manuscript notebooks of Jules Janssen detail prominencse observed from India on 19 August 1868, the day he discovered a method of obtaining the spectrum of the chromosphere, and where he detected a previously unknown gas, later named 'helium'.

Moon hides the edge of the photosphere, the chromosphere appears as an irregular fringe, partly caused by its structure, and partly because of the lunar relief. The latter is responsible for a beautiful and spectacular phenomenon, that of Baily's Beads. The name goes back to 15 May 1836, when the British astronomer Francis Baily observed an annular eclipse of the Sun, and noticed irregularities on the lunar limb. He described them as like a necklace of beads of light, and gave the true explanation: sunlight is blocked by lunar mountains, but passes through the intervening valleys. Generalizing, the complete ring of Baily's Beads appears whenever the diameter of the Sun lies between the minimum diameter of the Moon (corresponding to the bottom of the valleys) and its maximum diameter (corresponding to the mountain peaks). Baily's Beads were made a particularly great impression on the crowds that gathered at Saint-Germain, near Paris, for the annular eclipse of 17 April 1912.

During an eclipse, the Moon, which continues to move, takes a few seconds to cover the chromospheric fringe at the eastern limb of the Sun before revealing that on the western limb, at the end of totality. Astronomers thus have about ten seconds at the beginning and ten seconds at the end of an eclipse to photograph the upper chromosphere. To capture the lower chromosphere, which lies between 2000 and 3000 km above the photosphere, the time available is even shorter. It requires a flash spectrum, using a slitless spectrograph – the chromosphere being sufficiently narrow to act as a slit itself.

As for the corona, spectroscopic analysis also produces surprises. During the eclipse of 1869, Harkness and C.A. Young discovered a spectral line in the green region. They attributed it to an unknown element, called 'coronium'. But observations made by Jules Janssen in Hindustan in 1871 and from the Pacific in 1883, identified this line as that of a well-known element: neutral hydrogen.

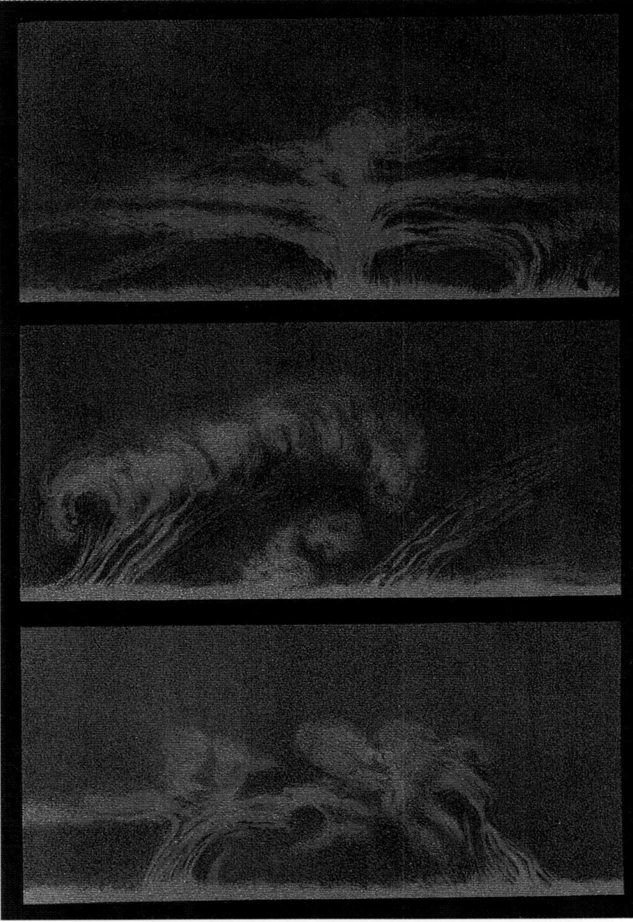

■ These solar prominences were observed from Palermo by P. Tacchini. *Top*, prominences observed 18 July 1860. *Centre*, the mixed, filamentary, and nebulous prominence dates from 22 December 1871, and bottom, the solar 'rain' corresponds to the period between 2 September 1871 and 7 July 1872.

Throughout his life, Janssen showed a remarkable tenacity in observing eclipses. In 1870, when Paris was besieged by the Germans, he escaped in a balloon to observe the eclipse of 22 December. When he arrived safe and sound at Oran, in Algeria, where he went ashore to observe the eclipse, clouds completely covered the sky!

The artificial eclipse: Lyot's coronograph

Because of the extreme brevity of total solar eclipses (just a few minutes), scientists thoughts soon turned to observing the corona outside eclipses. But the difficulty is considerable. Not only is the brightness of the corona in normal light just one millionth of that of the photosphere, but the sky in the vicinity of the Sun's disk is itself ten to one hundred times as bright as the corona, because of scattering of sunlight in the Earth's atmosphere.

In 1931, the French astronomer Bernard Lyot invented an instrument that enabled one to create artificial eclipses at will, to observe the corona in daylight and continuously. To do this, it is not sufficient to use just a black disk that exactly masks the photosphere. His coronograph also suppresses the stray light scattered by the optical elements in the instrument. It enabled Lyot to carry out daily visual, photographic, and spectroscopic observations of the inner corona.

In 1933, Lyot perfected his

instrument by adding a monochromatic filter that selected the appropriate radiation for different layers of the Sun. He was thus able to obtain photographs of the Sun at three important wavelengths, including the Hα (H-alpha) line of hydrogen. With this equipment, it is possible to make a film of prominences. From 1935, Lyot obtained some magnificent films, such as one called *Flames on the Sun*, showing the jerky movement of prominences and the fireworks of eruptions.

The author Claude Roy described the discovery in these words: 'Thanks to the coronograph that he [Professor Lyot] has invented, I have seen, erupting from the Sun's corona, and from its outer surface, the great sheaves of solar eruptions, those feathery plumes of gas in the act of fusion, whose vigour is measured [...] in hundreds of thousands of kilometres.'

The jewels in the crown

Many observatories around the world now have coronographs and spectroheliographs. Some sites, such as the Pic du Midi in France, Pico de Teide in the Canary Islands, and Sacramento Peak in New-Mexico, have even specialized in the study and surveillance of the Sun. Nevertheless, whatever the quality of the coronographs, only the inner regions of the corona are detectable outside eclipses. A total eclipse, which is the best of all coronographs, also allows observation of the outer corona, especially if one uses instruments mounted on aircraft that fly at high altitudes, where the rarified atmosphere is transparent and scatters very little of the sunlight. This is why, for 30 June 1973, which was to be a very long eclipse – 7 minutes 4 seconds – a French team led by Pierre Léna and Serge Koutchmy suggested that a suitable flying observatory would be – Concorde 001. The aircraft took off from the Canaries, caught up with the eclipse as it sped across Africa, and, flying at 2200 km/h, remained within the Moon's shadow cone for 74 minutes, while ten scientists studied the corona through portholes that had been specially polished for the occasion. Despite the duration and the altitude record – 17 000 m – of the observations, atmospheric turbulence, partly caused by the aircraft's supersonic shockwave, did limit the scientific value of this spectacular mission.

In general, before the advent of space missions, like Skylab in the 1970s and, above all, Soho from 1995 onward, astrophyicists wanting to record spectra and images of the solar corona were forced to travel, with all their scientific equipment and baggage, to places where eclipses were to take place. On 11 July 1991, however, the Sun and Moon offered them a fine present by visiting their own backyard, namely one of the most important observatories in the world, which is located on the top of Mauna Kea, on the island of Hawaii. This international site, perched at an altitude of 4200 m, is equipped with extremely powerful telescopes, whose sole fault is that they were not designed for observing the Sun! As a result, the scientists had to adapt instruments designed to observe galaxies one hundred million times as faint as the faintest stars visible to the naked eye, so that they could observe our star as it was eclipsed. A whole host of precautions had to be taken to protect the telescopes. The most powerful of them, the Canada-France-Hawaii Telescope, has a mirror 3.6 m in diameter. The slightest ray of light from the photosphere that struck this mirror would turn it into a solar furnace. So, a perfectly opaque mask, with just two minute pinholes, which enabled the Sun to be located and followed without danger, was placed in front of the mirror until just a few seconds before totality. At the exact moment when the last ray of light from the Sun disappeared, the mask was removed. For the four minutes that the eclipse lasted in Hawaii, the CFH Telescope was thus able to provide its full optical power. Never had such a powerful instrument been used to study the Sun. This time, the gamble paid off for the international team using the telescope, led by Serge Koutchmy from the Institut d'Astrophysique in Paris: the mirror's exceptional optics enabled them to photograph extremely fine structures, less than 300 km across, within the corona. This was a surprise to the physicists, who had never imagined that such small structures could survive in a medium that was so tenuous, at such a high temperature, and moving so rapidly. These discrete coronal threads, as Koutchmy has called them, have a very short lifetime of about one minute. During this memorable observation on 11 July 1991, Serge Koutchmy's team were also able to capture a solar structure, 1500 km in diameter, that had been completely unknown until then, and which he subsequently called a 'plasmoid'. It is a bubble of cold plasma, which originates

■ It was during the eclipse of 28 July 1851, here drawn by A.L. Busch, that the corona and prominences were photographed for the very first time, and definitely linked to the solar atmosphere.

■ At the eclipse of 18 July 1860, observed from Spain, Warren de la Rue drew the solar

at the surface of the Sun, and rises into the corona, before dissipating. A high-resolution film obtained with the CFH, centred on a moving plasmoid, shows the latter becoming distorted and breaking up. The movement of the plasmoid is not radial to the Sun, but followed the local magnetic field.

In a dome adjacent to the Franco-Canadian telescope, another team observed the eclipse with the University of Hawaii's 2.2 m telescope. Here again, the researchers had the use of an extremely powerful telescope to study the Sun. They put it to good use during the four minutes of totality by making a high-resolution film of a giant solar prominence, itself located, before the start of the eclipse, by the coronograph at a neighbouring solar observatory on Mauna Loa. Without this valuable contribution, the astronomers would never have had the time, with their large telescope, to 'tour the Sun' to find an interesting feature – the field of view of a telescope 2.2 metres in diameter is actually very tiny. Once the eclipse was over, computer enhancement of the original images, designed to improve their resolution, was applied to the film. The astronomers used the lunar limb, which should be extremely sharp, as a reference. They were thus able to partially correct for atmospheric turbulence that distorted the images as they were being taken. This spectacular film led to a better understanding of the fine structure of solar prominences, which had never been observed under such conditions.

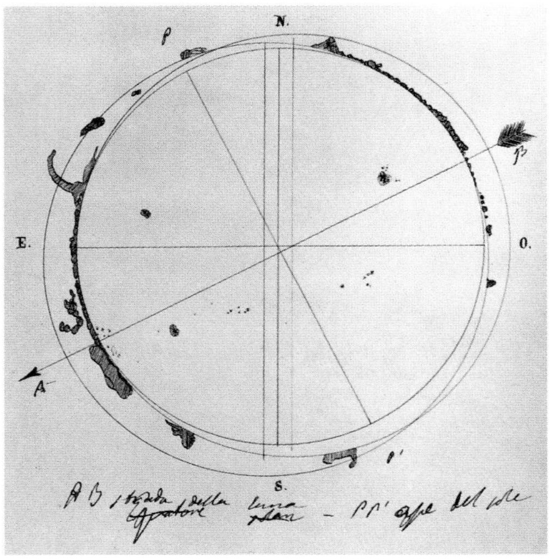

The shape of the corona varies from eclipse to eclipse, depending on the Sun's magnetic activity. This follows an 11-year cycle, which is particularly well shown by the appearance, migration, and disappearance of dark sunspots on the photosphere. At solar maximum – corresponding to a maximum number of sunspots – the corona is regular, almost

■ Angelo Secchi also made drawings and took daguerreotypes, but from a site 500 km away from De la Rue. Comparing their results, they deduced that prominences were 150 million kilometres away, and were definitely part of the Sun.

PROMINENCES (HERE REPRODUCED FROM *LE CIEL* BY GUILLEMIN) AND ALSO TOOK SOME DAGUERREOTYPES.

circular, in shape. At solar minimum, its shape is very irregular, and takes the form of extremely extensive equatorial streamers, which sometimes stretch out as far as twelve solar diameters, and which give the Sun a 'winged' appearance (*see* Chapter 2).

The existence of the Sun's overall magnetic field, which determines the shape of the corona was suggested in 1889 by Bigelow, and confirmed in 1891 by A. Schuster. As the latter said, the shape of the corona around minimum is 'very similar to what one would find if the Sun were a magnet, discharging negative electricity around the poles'.

Since it has become possible to observe the corona thanks to telescopes in space, scientific interest in eclipses has diminished. Outside the Earth's atmosphere, in fact, the sky background is completely dark, atmospheric scattering has disappeared, and the corona is distinctly seen. Spaceborne experiments, which have extended observations of the corona out to thirty solar radii, i.e., 20 million kilometres from the Sun's limb, have confirmed Schuster's intuition. They have also allowed us to define the physical properties of the transition zone between the Sun's atmosphere and the interplanetary medium.

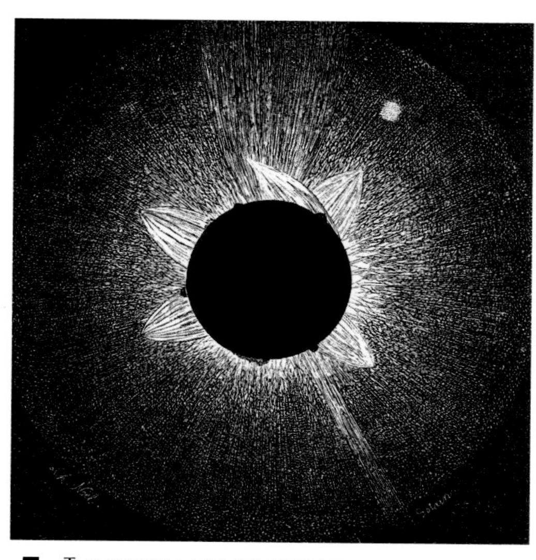

■ THE CORONA AND ITS STREAMERS WERE SPECTACULAR DURING THE ECLIPSE OF 7 SEPTEMBER 1858, OBSERVED FROM PARANAGUA BY THE ASTRONOMER EMMANUEL LIAIS, THEN ON A SCIENTIFIC MISSION TO BRAZIL, AND THE AUTHOR OF AN ASTONISHING BOOK, *TROPICAL NATURE AND HEAVENLY SPACE*.

The European spaceprobe Soho, launched on 2 December 1995, has been working between the Sun and the Earth, at 1.5 million kilometres away from the latter, where there is a point of gravitational equilibrium between the two bodies. Equipped with 12 instruments designed to study the solar wind, the Sun's atmosphere, and the body of the Sun itself, in different bands in the visible and ultraviolet regions of the spectrum, Soho has made numerous discoveries. These have included the existence of broad currents within the Sun's interior near the poles, and also allowed us to understand what had been the greatest mystery of solar

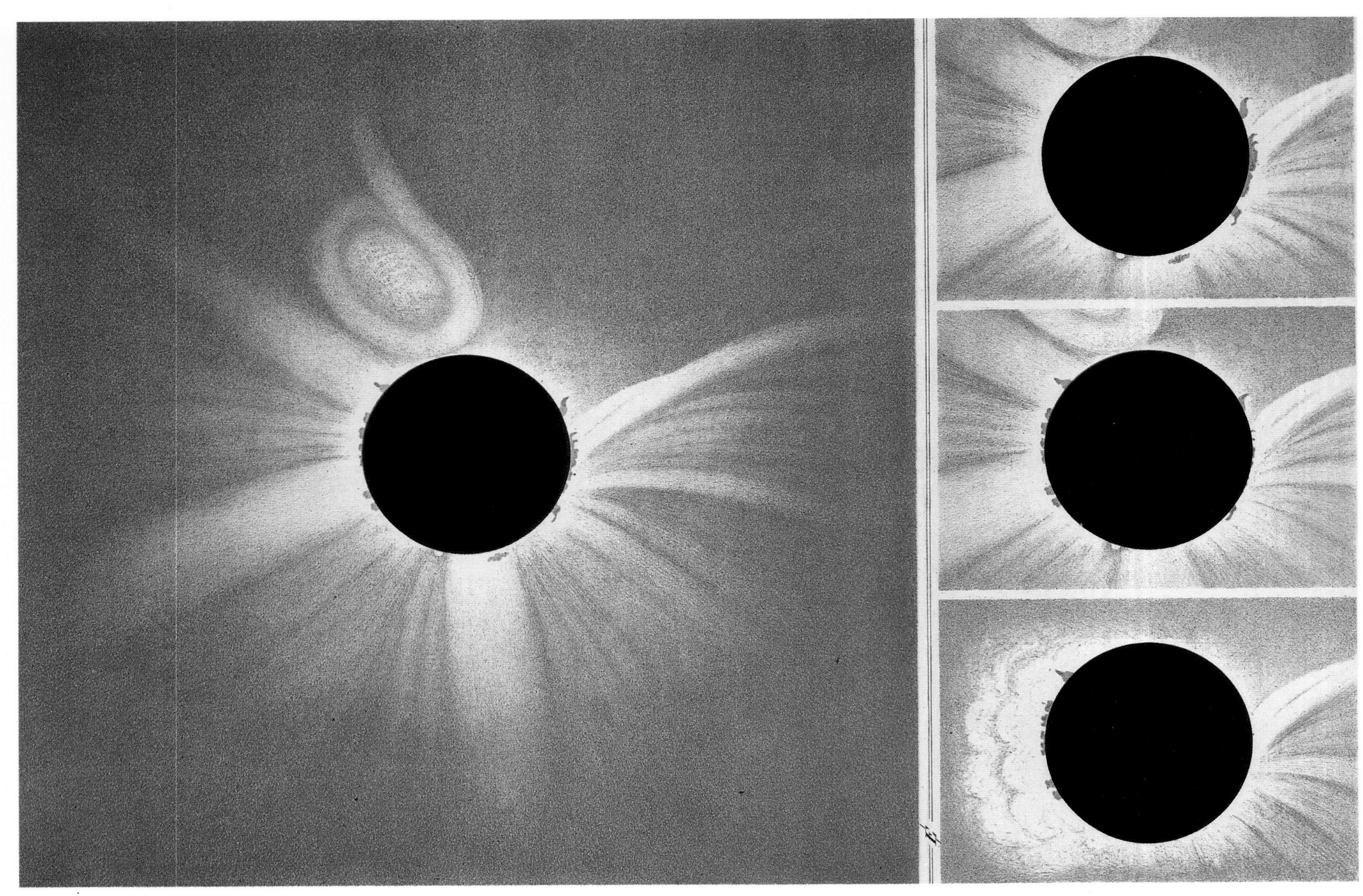

■ These magnificent drawings, preserved at Paris Observatory, faithfully reproduce the appearance of the corona and prominences observed by A. Secchi from Torreblanca, Spain, at the total eclipse of 18 July 1860.

physics until now: why are the temperatures found in the Sun's outer atmosphere so high? While the temperature at the Sun's surface is around 5000°C, it climbs to 2 million degrees above 10 000 km. Scientists have never been able to account for this extraordinary heating of the corona. A mechanical explanation was suggested in 1942 by Martin Schwarzschild, who had detected the extreme agitation of the solar surface at kilometre wavelengths: material was flowing upwards at velocities that reached hundreds of kilometres per second. The currents and the immense shockwaves that they produced were believed to explain the heating, but only in part, because the amount of energy involved remained insufficient.

Soho revealed a new aspect of solar magnetism: the dynamo effect. This phenomenon explains the origin of the magnetic fields through the rotational and convective motions in the fluid, conductive interior. Its role in the Sun nowadays appears to be much greater than previously thought, because it may also explain the heating of the corona. A continually changing 'magnetic carpet' of lines of force, which creates magnetic currents incomparably greater than on Earth, is woven throughout the Sun and its atmosphere. Imagine an electrical network that is constructed by connecting powerful generators through good conductors: such a network would heat up. Similarly, the Sun's magnetic network causes the strongest heating just where the loops are closed, that is, at altitude.

Eclipses and the curvature of space

Eclipses are a phenomenon that involves the Earth, the Moon and the Sun: three bodies that occupy a derisively small region of space when compared with the immensity of the Galaxy, never mind the universe as a whole, with its hundreds of billions of galaxies. Yet eclipses have allowed us to establish, on a true experimental basis, the fundamental theory about the nature of space, time, matter, and light: general relativity. This has literally changed our vision of the cosmos, notably by leading to the concept of a universe that is expanding from an initial Big Bang.

■ These fireworks were immortalized by a series of photographs begun by Bernard Lyot at the Pic du Midi Observatory at 9 o'clock in the morning of 12 June 1937. Lyot later made them into an excellent film.

In 1905, Albert Einstein published three papers, each worthy of a Nobel Prize. One explained Brownian motion, another the photoelectric effect, and the third advanced the theory of Special Relativity. The success of these papers transformed an obscure employee in the Patent Office in Berne into one of the foremost among the small group of physicists. This phenomenal start was followed by other major advances, but what was by far the most influential of his work was the theory of General Relativity, which describes the distortion of space-time by a gravitational field. In 1911, when he was working at the University of Prague, Einstein made the first calculations, but he did not take any astronomical data into account. It was the German astronomer Erwin Findlay Freundlich who convinced him of the importance of observations that would test his theory of gravitation, notably the deviation of light rays near the Sun. According to General Relativity, the curvature of space near a massive body should cause rays of light passing nearby to curve inwards. A star seen at the edge of the Sun's disk at the time of eclipse, would thus appear to be shifted through a certain distance relative to its normal position.

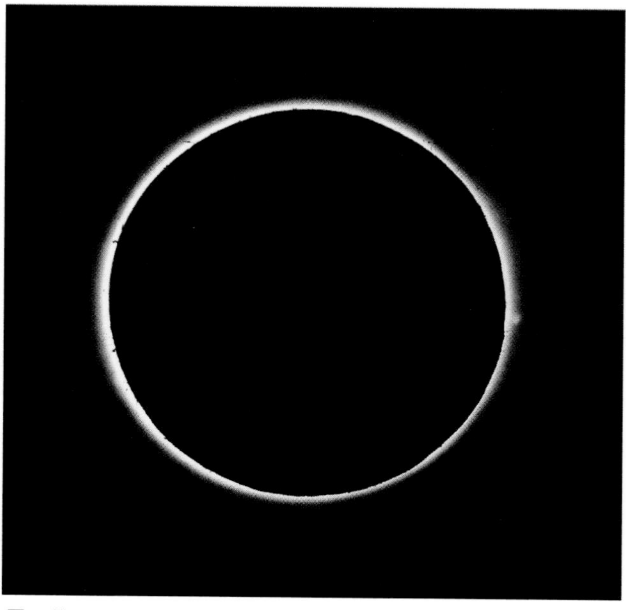

■ Thanks to his invention of the coronograph, capable of creating an artificial eclipse at will, Bernard Lyot was able to photograph the solar corona in full daylight and continuously. This is one of the very first photographs taken using this technique in 1931.

A useful image of curved space-time, and of its influence on the motion of light and matter, is that of an elastic sheet. Imagine a piece of space-time reduced to two dimensions, in the form of an elastic sheet. In the absence of any object on top of it, the sheet remains perfectly flat. If one places a ball on it, the sheet deforms, and creates a hollow around the ball. The heavier the ball, the deeper the hollow. If, in addition, the sheet is actually a network, the threads in the net allow us to visualize the paths that rays of light will follow. They form a perfectly rectangular array in the absence of gravitation, but deviate from a straight line in the presence of any mass, following the warped shape of the sheet. This deviation of the rays of light passing close to the Sun is one of the specific effects of General Relativity predicted by Einstein that may be verified experimentally. It should be revealed by a slight displacement in the position of stars when the Sun is close to their line of sight. Naturally, no one could expect to observe such a phenomenon except during a total eclipse of the Sun.

A German expedition to the Crimea was arranged for the 21 August 1914, the date of a solar eclipse. But that was not a judicious time to be a German on Russian territory. The team was imprisoned for a month until an exchange could be arranged, and no scientific photographs were taken. Some historians of science have noted that it was perhaps just as well: the calculation of the deflection of light that Einstein advanced in 1911 was in error by a factor of 2, and the Crimean observations would have contradicted his prediction. Knowing Einstein's personality, however, we may well imagine that such a reverse would not have discouraged him.

However that may be, in the meantime Einstein had realized that his theory was inaccurate. He finally succeeded in formulating it in 1916, in particular calculating that the true value of the deviation of rays of light at the Sun's limb would be 1.74 seconds of arc.

During the First World War, few scientific ideas were exchanged between the Germans and the Allies. When the British astronomer Arthur Eddington received Einstein's paper on General Relativity 'through the grapevine', he immediately realised its importance.

A total solar eclipse was due to occur on 29 May 1919, along a line passing through Brazil, the Atlantic, and West Africa. Eddington and his colleague Frank Dyson organised an expedition with the aim of detecting the effect predicted by Einstein's theory. The moment when Eddington proposed the expedition corresponded to the peak of the war, when prospects for the Allies appeared at their darkest. A British test of a German theory would be a major proof of the internationalism of science. But Eddington was a Quaker, and thus a confirmed pacifist, and he gained his ends.

■ From this historical photograph, taken during the eclipse of 29 May 1919, the deviation of light rays from stars lying in the background to the Sun was measured for the first time, and found to agree with the theory of General Relativity as proposed by Albert Einstein in 1915.

To better guard against the risk of bad weather, Eddington and Dyson chose two sites: one on the island of Principe, on the West Coast of Africa, which was where

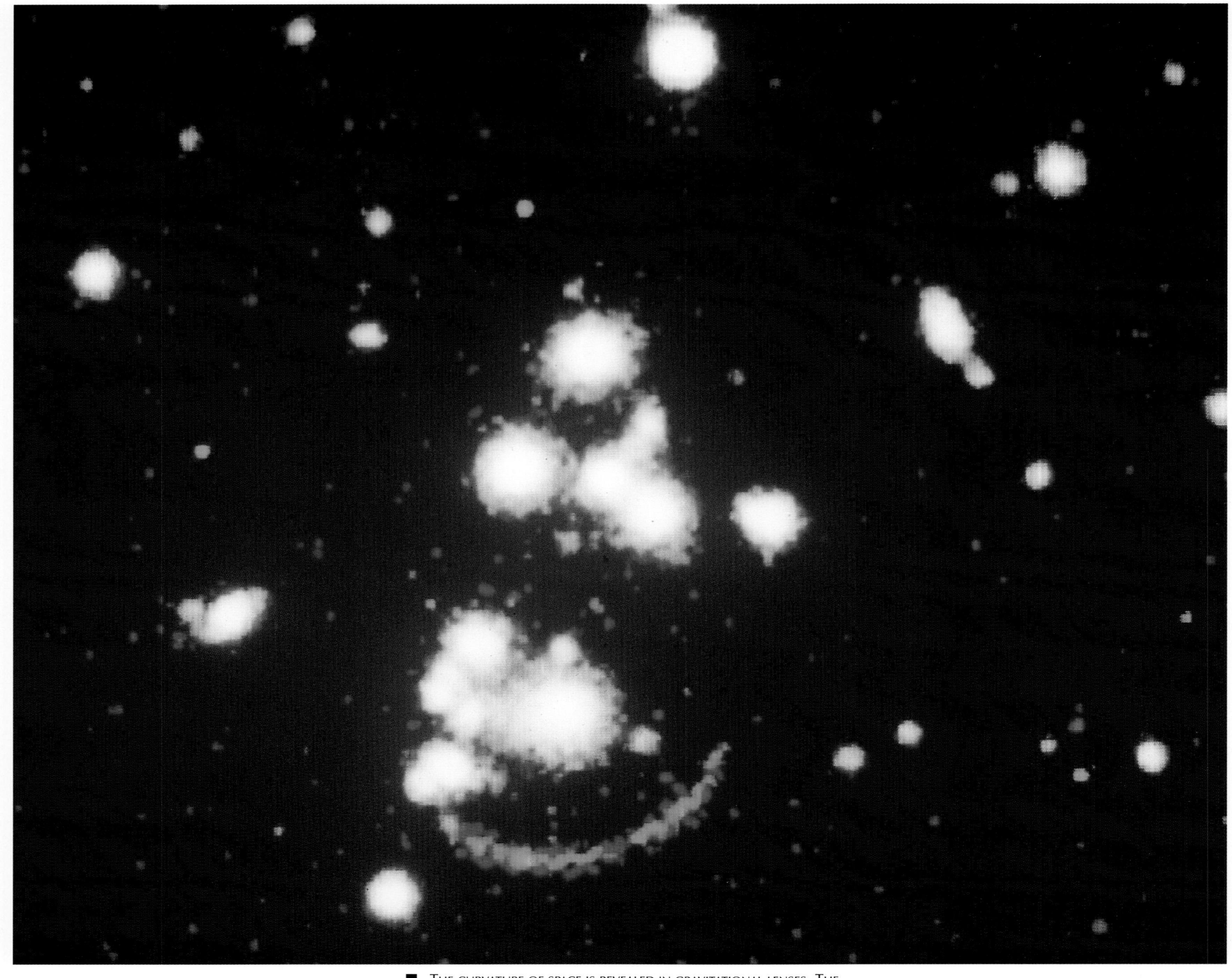

■ The curvature of space is revealed in gravitational lenses. The luminous arc is the distorted image of a distant galaxy, lying far behind the cluster CL 2244–02.

Eddington went, and the other at Sobral, in Brazil. At Principe, on the day of the eclipse, it started to rain heavily; the rain ceased at midday, but the Sun did not appear until after first contact. Nevertheless, Eddington took 16 photographs through scattered clouds. Six months later, he returned to the site to take the same field of view, in the absence of the Sun, i.e., at night, so that the apparent positions of the stars could be compared. The results confirmed Einstein's theory, giving values of 1.98 + 0.30 arc-seconds and 1.61 + 0.30 arc-seconds. They were announced on 6 November 1919 at a memorable meeting of the Royal Astronomical Society. The audience fully realized that they were witnessing a turning point in the history of physics. The international press trumpeted relativity's success, and Einstein became the symbol of scientific genius.

■ Over a period of forty years, Jules Janssen observed a large number of eclipses. Here, the expedition of 30 August 1905 is installing its instruments at Alcosebre in Spain. The spectrographs allowed the composition of the solar corona to be determined.

Einstein was not surprised. He later told the tale of how he was in the company of Max Planck, the German physicist who played a pioneer role in the theory of quantum mechanics. 'He was one of the most intelligent people I have ever known; but during the 1919 eclipse, he stayed up all night to see if it would confirm the deviation of light by the Sun's gravitational field. If he had really understood the way in which General Relativity explains the equivalence of inertial mass and gravitational mass, he would have gone to bed like me!'

Albert Einstein obtained the Nobel Prize for Physics in 1921, but for his explanation of the photoelectric effect, not for his brilliant theory of relativity! ■

■ The bluish objects are multiple images of the same galaxy. This gravitational-lens effect is created by the cluster of galaxies called CL 0024+1654, lying in the centre of the image, and yellow in colour.

■ At Mauna Kea Observatory, on Hawaii, the 3.6-m France-Canada-Hawaii Telescope was turned, with extensive precautions, towards the Sun as it was hidden during the total eclipse of 11 July 1991.

11 August 1999: the last eclipse of the millennium

■ On 11 August 1999, at 10:30 Universal Time, for the first time in history, people were able to observe a total eclipse of the Sun from space. Glued to the porthole on the *Mir* Space Station, where he had been living for nearly six months, the astronaut Jean-Pierre Haigneré followed the track of the Moon's shadow racing at around 3000 km/h across the clouds that covered northern Europe.

■ WERE THE MEGALITHS OF STONEHENGE, THREE THOUSAND YEARS AGO, A DEVICE FOR CALCULATING ECLIPSES? ON 11 AUGUST 1999, THE MOON'S SHADOW PLUNGED THE STONES INTO DARKNESS.

It is a dark night; as dark as it is possible to imagine. The sky is like ebony, pricked by ten thousand stars, as bright and cold as crystal. It is icy cold in the vacuum of space above the deserted landscape, which is illuminated by a faint ashen light, with indistinct shapes that loom in silhouette against the starry background. A vast dark plain, a few craters, gentle hills, and finally vast slopes that stretch to the horizon: this is the crater Hipparchus. And, up above, as wonderful and as blue as a drop of life, is the Earth...

On the Moon, on that 11 August 1999, at 9 o'clock in the morning, there was no one to admire the magnificent spectacle of the Full Earth. And what a spectacle that was! Here, night had lasted nearly two weeks, and the temperature had fallen to –170°C. High in the sky, not far from the zenith, the Earth remains stationary – because the Moon always turns the same face to our planet – but changes, because, like the Moon as seen from Earth, the latter, seen from its satellite, passes through all its phases: first New, then a crescent, then First Quarter, gibbous, Full, gibbous again, Last Quarter, waning crescent, and finally New once more.

That morning, if there had been any spectators on the slopes of the crater Hipparchus, they would have been able to admire the Earth at Full. A blinding object, nearly four times the size of the Moon as seen from Earth. Towards the west they would have been able to see the two Americas slowly rousing with the Sun. They would admire the blue Atlantic Ocean dotted with delicate whirls, blindingly white, of the depressions that were following one another across northern Europe, which was partly hidden below the clouds. Farther east, they would see the subtle ochre shades of the deserts on Asia, mingling with the varied greens of the great plains and forests of Siberia. Finally, and most strikingly, as clearly as in some geographical atlas, the silhouette of Africa against the blue background of the Atlantic and Indian Oceans, would dominate their view...

There is a dull, intolerable and inhuman rumbling. A sea of light, of radiation, at an unheard-of intensity. A maelstrom of nuclear fire, burning, exploding, vibrating, and howling without the slightest respite for 4.6 billion years. On the morning of 11 August, the Sun is full of seething energy. Every eleven years, following its cycle of magnetic activity, its surface is covered with dark spots, each an apparent abyss capable of engulfing the Earth several times over. Higher up, the two atmospheric layers – the chromosphere and the corona – which surround the photosphere, itself at a temperature of 5500°C, are also the site of violent disturbances. At the end of the millennium, the solar cycle is approaching its climax. That morning, right on the northeastern limb, a powerful magnetic fountain has been gushing forth for several hours, and masses of searing plasma

■ A meteorological miracle just a few tens of kilometres to the north of Paris: the clouds part, the very last ray of light from the Sun is just about to disappear, and the dark disk of the Moon is already visible. The eclipse is about to begin.

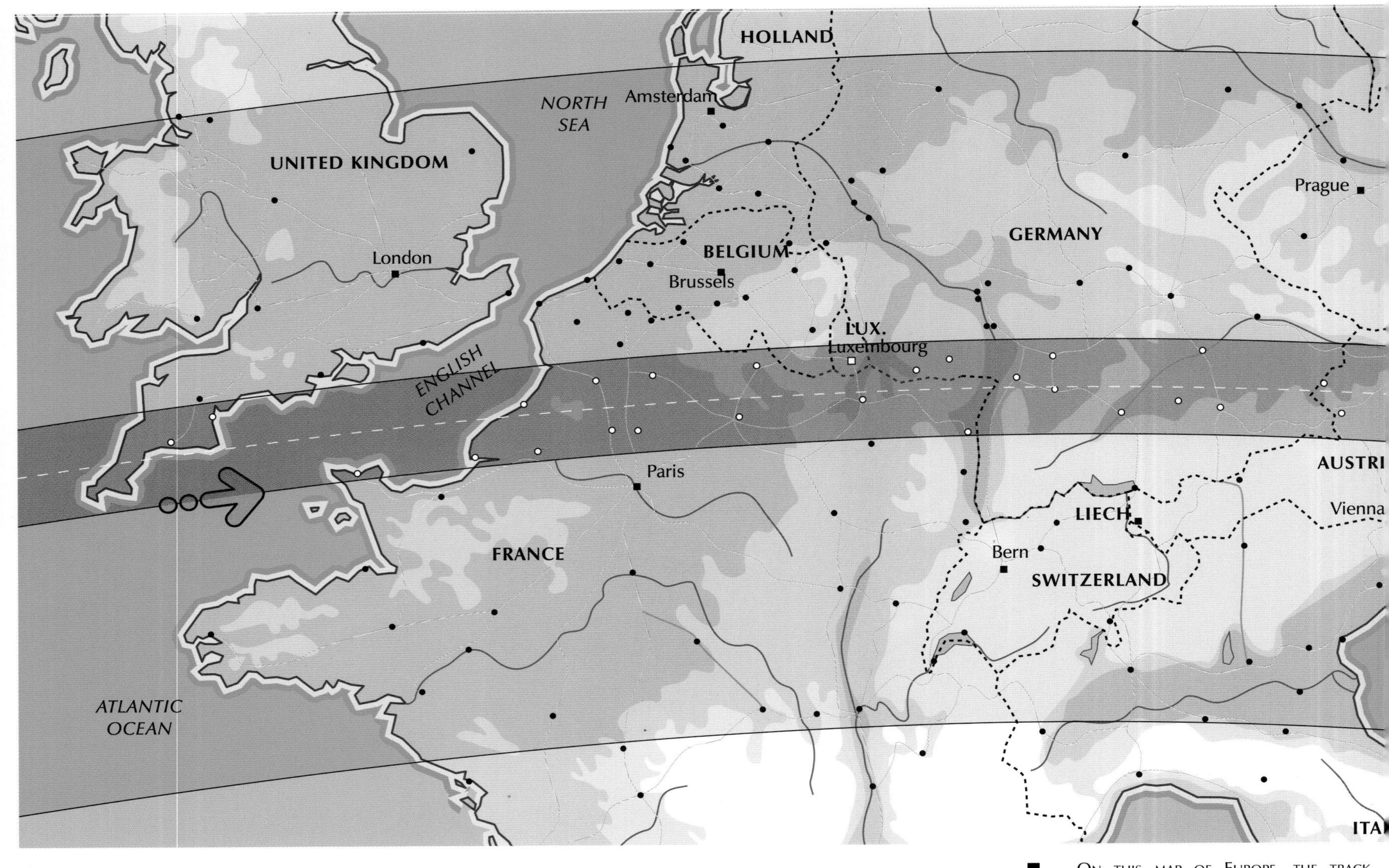

■ On this map of Europe, the track followed by the Moon's shadow is shown by the central dark band. On either side, the shaded zone indicates where the Sun was

are beginning to rise, leaving the dark, almost calm, and nearly cold well of a sunspot. In the spot, the gas at its surface is at a temperature of less than 4000°C, but the plume of hydrogen, which, upheld by the lines of magnetic force, has begun its rise towards the sky, has already exceeded 10 000°C. In a few minutes, the prominence, ejected by the incredible violence of the solar storm, will attain a velocity of 1000 km/s, force its way through the searing loops of the lower corona, and, inexorably, will escape from the powerful gravitational field of our star.

The actors are ready

On the morning of 11 August 1999, it is a very strange dawn that slowly grows off the shores of Nova Scotia and Newfoundland, in the North Atlantic. As with the slopes of the crater Hipparchus on the Moon, and the blinding, deadly clouds on the Sun, here also there are no witnesses of the drama that is about to unfold. A few humpback whales from the nearby Gulf of St Lawrence, blow for a moment at the surface, indifferently, before continuing their long journey towards the more pleasant waters of the Caribbean. In the sky, the stars disappear one by one, as if the dawn was near. Indeed, for a long moment, over the waves towards the eastern horizon, just as it does every morning, the flush of the dawn appears to precede the rising Sun. But then everything changes, and something extraordinary happens. Over in the east, slowly, the dawn fades and disappears, giving way to an incomprehensible, soft ashen darkness. One by one, the brightest stars, first Capella, then Procyon, Betelgeuse, and Sirius, shyly resume their places in the sky. Towards the horizon, the planets Saturn and Mercury reappear in their turn.

■ All great European cities came to a standstill between 10:00 and 12:00 Universal Time. Spontaneously, groups of people gathered to witness the event. Here, the crowd surrounds the Grande Arche, in Paris.

It is 9:30 Universal Time: the time shown by clocks at Greenwich in England, over there, far towards the east. Some 372 000 km away, on the Moon, an observer would notice something strange on Earth, like a veil of thin gauze, obscuring the North Atlantic. Much farther away, at a distance of some 150 million kilometres, on the blinding limb of the Sun, the prominence, swept up by a surge of energy, has slowly become detached from its star, and continues is free ascent into space.

It is 9:30 and 57 seconds. The Sun should have risen, but it is still night. On the horizon, it is like the

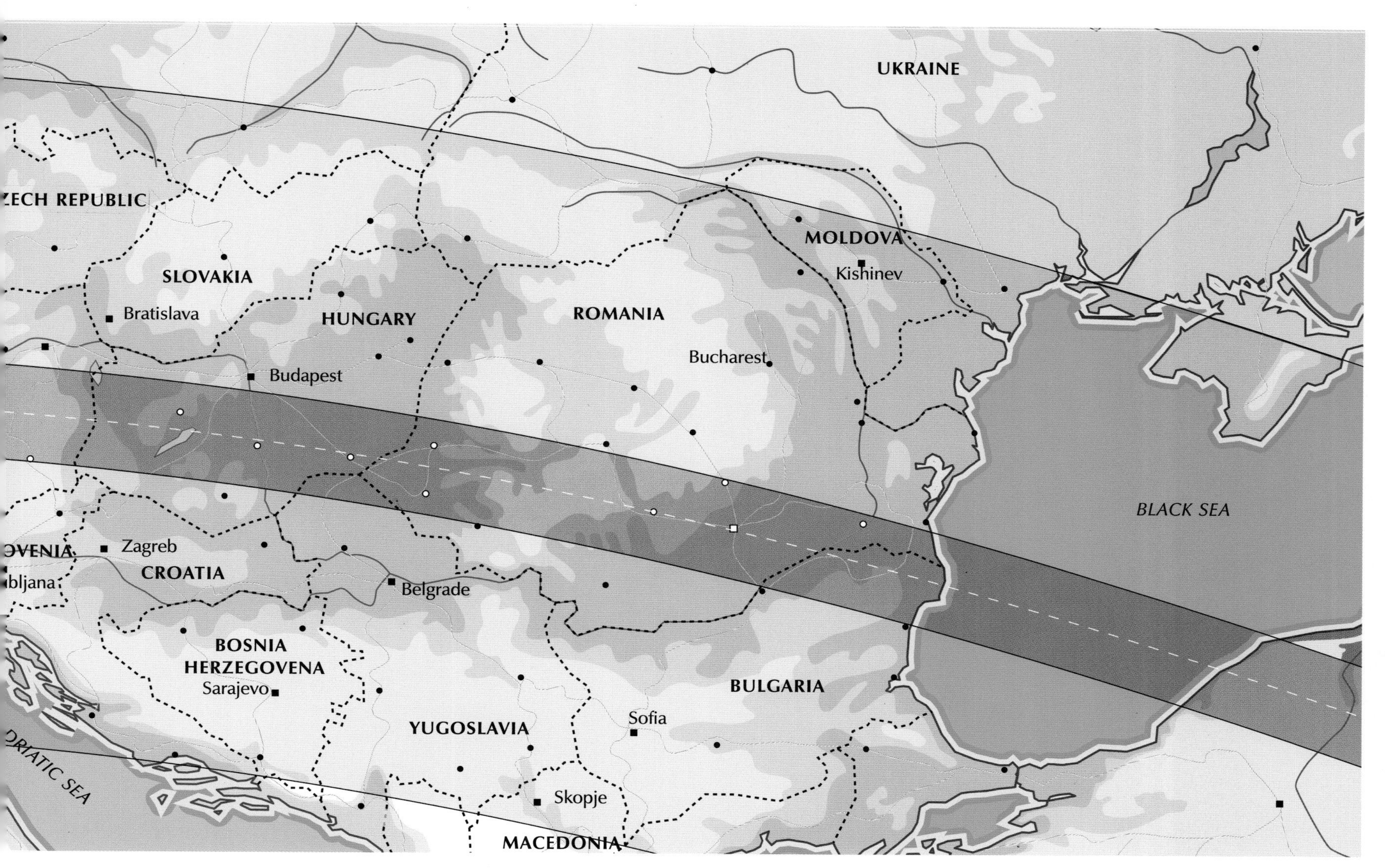

more than 90% eclipsed. The shadow of the Moon crossed Europe from one end to the other in about an hour.

scene at the end of the world: above the waves, a black, elongated disk, slowly rises through the mists of the early morning. The total eclipse has begun.

Let the spectacle begin

This planet had not seen a total solar eclipse since 26 February 1998. Twice since then, on 22 August 1998 and 16 February 1999, the Sun-Earth-Moon alignment had been perfect, but the Moon had been too far from the Earth to completely hide the Sun, and those eclipses, known as 'annular' ones, had not excited the crowds, nor mobilized every eclipse chaser. But this eclipse was exceptional. To begin with, it was the last of the millennium, the last in a cycle of 781 total eclipses that began with that of 19 September 1001. Then, this eclipse followed a rare path across the Earth, crossing some of the most densely populated countries, from northern Europe to India. Finally, it was the first eclipse that had occurred since the Earth had become a 'global village', interconnected from one pole to the other via the Internet, mobile phones, and television dishes. To put it another way, never before, in the whole history of humanity, had an eclipse of the Sun affected so many people, and never had an astronomical event been so glamorized by the media.

■ The magic of the eclipse of 11 August 1999 indiscriminately crossed the densest connurbations in Europe, Pakistan, and India, as well as remote deserts in Turkey and Iran. Here, the eclipse is seen from Hungary.

That morning, however, off the coasts of Nova Scotia and Newfoundland, there was no one in the front row to admire the black sunrise. To appreciate the spectacle, one would need to have been on the Moon: from the slopes of Hipparchus, one would have been able to see clearly the tiny black spot start on its journey across the Atlantic at more than 7000 km/h.

At the same time, in Europe, which the shadow of the Moon would soon reach, around four hundred million people turned their faces to the sky. The weather was not co-operating: one depression followed another, covering the south of England, the north of France, the south and west of Germany. Luckily, farther towards the east, conditions were far better: in Austria, Hungary, Romania, Bulgaria, on the shores of the Black Sea, and then, on the other side of the Bosphorus, in Turkey, the Sun, high in the sky, calmly awaited its appointment with the Moon.

France, which was shortly to be crossed from one side to the other

■ Reims remained one of the high spots for observing the eclipse. A crowd of ten thousand persons crammed the square in front of the magnificent Gothic cathedral, when, by lucky chance, a clearing in the clouds allowed the spectators to witness the grandest natural spectacle.

■ This magnificent photograph, taken with an astronomical telescope, shows the solar chromosphere and its ring of prominences. On the right, one of the 'flames' is breaking away from the Sun and about to fly off into space.

by the shadow of the Moon, and where the eclipse had occupied all the media for several weeks, was floundering in a sort of happy chaos. For eclipse chasers, the phenomenon took on a more-or-less sociological dimension, in the light of the sudden enthusiasm of their contemporaries for the finest sight in nature. They alone remembered that the event had been foreseen long ago: Camille Flammarion had announced it in his book *Astronomie populaire* as early as 1879! In fact, amateur astronomers the world over had long ago selected the best sites on the central line: all the hotels, all the camp sites, and all the rural cottages had been booked up for a year or more. On top of the cliffs at Étretat or Fécamp, around chateaux and cathedrals, on the shores of lakes, at the top of all the hills that dot the countryside from Normandy through Champagne as far as Alsace, they have been there, for several days, appraising the skies and checking the position of the Sun in the sky at time T on 'the Day'. Those who were unprepared, and who decided on the spur of the moment to 'go and see the eclipse' were not finding it much fun. At the Gare du Nord in Paris, all the trains were besieged, the autoroutes leading north and east were jammed, the main roads to the most desirable sites, such as the cathedrals of Laon or Reims, were saturated. At the same time, the same scenes were repeated more-or-less everywhere, around Plymouth, south of London, at Strasbourg, Stuttgart, Munich, and as far as Bucharest. Eclipse lovers, with their ears glued to their radios, desperately searched for any corner of the country where the weather might improve. Sophie had made her decision. A professional photographer, she had followed the advice of a friend who is an eclipse chaser, had checked the area the day before the eclipse, and had then set herself up, at 5 o'clock in the morning, with all her equipment, on the square in front of Reims Cathedral, before it was invaded by ten thousand people. She was right: her view of the eclipse was to be exceptional.

Waiting for the eclipse between the clouds

Christine and her friend Pascal, for their part, had decided to shun the crowds, the fireworks, and great gatherings. They had decided to stand right in the middle of a field, in Normandy. Christine was seeing her fourth total eclipse, and she had come to realise the extent to which, at a basic level, whatever the civilizations, whatever the cultures, eclipses always engendered the same fears, and the same fantasies. They too, reveal the ideas and anxieties of their time. Once upon a time, they reflected the particular relationship between Man and the gods. As far as 11 August 1999 was concerned, astrologers, divines, and gurus had all attempted to worry the masses with all sorts of apocalyptic predictions, although without any real success. Worries were elsewhere: millions of Europeans were worried, but about their health and that of their animals! Would the latter be stressed by the eclipse? In India, Christine had been saddened by the fact that the superstitious had hidden indoors during the eclipse of 24 October 1995; she discovered that in France, on 11 August 1999, the celestial event caused equal concern. Everyone seemed to have forgotten, or simply did not know, that there is almost one total eclipse a year – and that it does not bring the Earth to a standstill.

A few seconds to eleven o'clock (Summer Time). In southern England and northern France, a roar accompanies 'first contact', that precious moment when the dark limb of the Moon touches the Sun's limb. The weather is abominable, and generally, nothing at all is visible through protective spectacles. Paradoxically, the public observes the Sun with the naked eye or through ordinary sunglasses, it having been filtered naturally and cruelly by a thick layer of clouds!

At the time, Emmanuel is stuck in a traffic jam on the Autoroute du Nord, somewhere in the henceforth notorious 'zone of totality', some one hundred kilometres wide, where, and only where, the eclipse will be total. On either side of this narrow band, the material representation on Earth of the path of the Moon's shadow, the eclipse will be just partial. For him, the event has already started. He would never have imagined anything like it: on the motorway, every toll, every rest area, is invaded by drivers, who, in default of anywhere better, witness the spectacle alongside petrol pumps, from the cafeteria with its 'eclipse special' menu, or from the heavy-goods area. As more and more of the Sun is covered, traffic seizes up, as drivers stop so that they too can marvel at the wonderful spectacle.

■ At Jerusalem, the eclipse of 11 August was 80% partial. A group of Palestinians uses various methods of protection: a welding mask, special spectacles, and a simple piece of glass, smoked with a candle.

J - 143
AVANT L'AN 2000

■ An eclipse 143 days from the year 2000 excites the imagination! At Paris, night practically fell at 12:23 (Summer Time), but a very narrow crescent of the Sun – less than 1% of its diameter remained uncovered by the Moon! – and, above all, the clouds, prevented Parisians from seeing the prominences and the corona around our star.

■ The Normandy countryside slowly fades into darkness. This fine series of photographs, all taken with the same exposure, shows the very last seconds before the eclipse. Below the

In their field, surrounded by nature, Christine and Pascal feel the temperature slowly falling. But the landscape has not changed. Beneath the shade of the apple trees, the herd of cows continues to graze contentedly. The Sun is more than half covered, but the light does not seem to have fallen. In fact, the decrease in light is definitely real, is not detected at first, for the simple reason that the pupil of the eye dilates more and more to compensate as the Moon advances over the disk of the Sun. This is why many partial eclipses – let us say those where less than 75% of the Sun is hidden – may pass completely unnoticed by spectators who have not been informed of the forthcoming celestial encounter.

Midday, and Europe is plunged into night

Midday. By now, it is impossible to escape from the eclipse. The Sun, up there in the sky, is still blinding, but something has definitely changed in the landscape. Christine recognizes sensations that she has felt before, unique to total eclipses, and to them alone. An increased freshness, the light from the Sun – at which she casts a knowing eye from time to time – is less insistent, and then there is the blue of the sky that is inexorably darkening. Very soon, but still in sunlight, Pascal spots Venus. That's the signal; Christine, whose heart starts to pound, knows that everything is going to happen too quickly, that the landscape will change from minute to minute, that it is too late to change the area framed by her camera or its settings. The long grass in the meadow takes on an indefinable colour, silvery, electric, and unreal. High above, the Sun is fading, and in the deepening blue sky, Venus shines with incredible brilliance. It is twenty-two minutes past twelve.

■ At Truro, in England, the members of the renowned British Astronomical Association adjust their telescopes and lenses before the great moment, awaited in Great Britain ever since 1927.

High above in the stratosphere, at an altitude of 18 000 m, for a short time, three supersonic Concordes follow the shadow, which is now moving at nearly 3000 km/h. At an even greater height, between the Earth and the Moon, three men witness a total eclipse from the void of space, for the very first time. The astronauts Sergei Avdeiev, Viktor Afanassiev and Jean-Pierre Haigneré, who have been on board the Space Station Mir for six months, glued to the

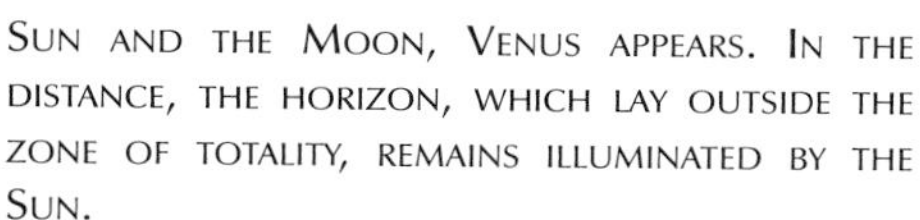

Sun and the Moon, Venus appears. In the distance, the horizon, which lay outside the zone of totality, remains illuminated by the Sun.

porthole that looks down on the Earth, watch the shadow of the Moon 'like a black finger touching the Earth'.

An hour earlier, Emmanuel, in his motorway parking area, and well away from the lampposts, which had come on automatically, thinking that sunset had arrived, feverishly unpacked his astronomical telescope and pointed it at the Sun. He awaits the total phase before uncovering the precious objective. It is his first eclipse, he feels sick, and is shaking.

A black finger touching the Earth

Suddenly, for the eclipse chasers, everything begins again. Suddenly, for those who have never seen the Moon hide the Sun, they are starstruck. A black veil has silently arrived from the west and covered the sky. Instantly, the Sun has vanished. To her surprise, Christine sees a solar corona like none she has ever seen before: circular, compact, and brilliant, it glows around a Sun that seems in a black mood. Some tens of kilometres away, Emmanuel stares open-mouthed at the spectacle revealed by his telescope: at the striped, luminous arches that stretch out from the hidden Sun, at the almost circular, harsh, and tangled corona, typical of solar maximum, and a contrast to the coronae seen at solar minimum, which are softer, elongated, and with a serene beauty. Above all, however, Emmanuel's gaze is fixed on the edge of the Moon, where a dozen solar prominences are visible. Their colour, a reddish-pink that has no equal, arises from the ionization of atomic hydrogen, and is perfectly pure. Through his telescope, Emmanuel, transfixed, unbelievingly stares at their loops which arch high above the narrow pink trace of the chromosphere. On the northeastern limb of the hidden star, the young amateur astronomer witnesses an astonishing spectacle, which rarely occurs during an eclipse: a solar prominence has become detached from the incandescent sphere and is probably escaping at some immense velocity. Seen from here, however, these plumes on the Sun appear perfectly stationary, as if fixed in space and time.

■ That Wednesday morning, the weather conditions were abominable at Fécamp, where tens of thousands of people had gathered. The clouds cleared just at the moment of totality.

In this chaos of images and emotions, among the other spectators who are shouting, applauding, and even crying, Emmanuel,

■ At Istanbul, in Turkey, the eclipse was only partial. The photographer made a double exposure to obtain this magnificent photograph of the dome of Hagia Sofia, dominated by a crescent Sun.

bewildered, tells himself that he has forgotten something important. But it is too late: at the western limb of the Moon, the solar chromosphere appears, and between the mountains at the lunar limb, the very first ray of the Sun appears. The eclipse is over.

Alone in their field, Pascal and Christine are deeply moved, and don't say a word. After a long moment, seeing the cows continuing to graze imperturbably, and remembering that they had not even cast an eye on the sky during the eclipse, they decide that eclipses undoubtedly do not always cause the same stereotypical behaviour. But Christine is already thinking of the 'next one', on 21 June 2001, just 681 days to wait...

As far as Sophie is concerned, she was lucky. On the forecourt of Reims Cathedral, black with people, and perfectly framed in her equipment, the total eclipse appeared for a few seconds between the clouds. Around her, thousands of people are congratulating one another, kissing, and shouting 'Encore, encore, encore' in a cacophonous euphoria.

The Sun has returned. It both illuminates and warms the verges of the motorways where, one by one, in a happy and contented disorder, cars continue on their way. Emmanuel quietly packs up his equipment, when, suddenly, he remembers. But of course, it is too late! He had decided that, during the total eclipse, he would observe the Moon, which everyone believes to be black during an eclipse, but which is, in reality, faintly illuminated by Earthlight. Emmanuel had not had time to notice this faint ashen light, he was so taken by the dreamlike vision of the solar prominences, which seemed, simultaneously, to be surging out from the Sun, and to be frozen in space. He

■ In the north of France, the spectators were frequently able to admire the partial phases of the eclipse with the naked eye. The natural filter of the clouds offered a magnificent view. Here, the Sun is peeping through clouds above the pinnacles of Rouen Cathedral.

■ The eclipse's farewell to Europe. In this magnificent photograph taken with a wide-angle lens, the shadow of the Moon racing towards the east is partially

also, starts to think of the 'next one', 21 June 2001, just 681 days to wait...

From the slopes of Hipparchus, on the Moon, the show goes on. The tiny black spot – 1′ in diameter, as astronomers would say – continues on its way across the blue planet. After England and France, it crosses Germany, Austria, Hungary, Romania, Bulgaria, Turkey, Iraq, and soon arrives at Iran, where a crystal-clear sky has attracted a number of both amateur and professional astronomers. At Isfahan, Jean-Pierre and Jean-Marc, both astrophysicists, see their second total eclipse together, and have to bow to the evidence: they, too, have become eclipse chasers.

When the shadow of the Moon crosses Pakistan, the Sun has begun its long decline towards the horizon, offering wonderful images for photographers seeking a cosmic viewpoint. From there on, the eclipsed Sun will no longer be seen by anyone on the ground, because the Moon's shadow is now running across monsoon clouds. At Karachi, and then in India, between Baroda and Chandrapur, millions of people are plunged into a dark, disturbing night. And when, later, the shadow of the Moon disappears off the eastern limb of the planet, after its mad rush over more than 14 000 km in just 3 hours 7 minutes, and the Sun sets in the Bay of Bengal, more

VISIBLE. IN THE DISTANCE, BEHIND A LINE OF CLOUDS, THE COUNTRYSIDE ILLUMINATED BY THE SUN WILL SOON FADE INTO NIGHT.

than one tenth of the world's population, some six hundred million people, have experienced the magic of the black Sun.

On the Moon, the Earth, still in the same place high in the sky over the ramparts of Hipparchus, begins to slowly wane. Bit by bit, its seas and continents disappear, plunging the blue planet into night; first gibbous, then appearing as a magnificent crescent that gets narrower and narrower. Here as well the eclipse is over, but everything will begin again in twenty-three lunations.

In two or three days, after a trip of some one hundred and fifty million kilometres, the ionized particles ejected into space by the powerful eruption of 11 August, henceforth lost to the Sun, and carried beyond the corona by the solar wind, encounter the Earth. Captured by our planet's powerful magnetic field, they electrify the upper atmosphere, painting sky above the North Atlantic with the gauzy, changing veils of a magnificent aurora. Perhaps, once again, there was no one to admire this curtain of light from the depths of space, that floated between the Great Bear and the Pole Star, there, off the coasts of Nova Scotia and Newfoundland, where, on 11 August 1999, the last total eclipse of the millennium began at dawn. ■

Atlas of eclipses of the Sun and Moon

■ FOR SEVERAL YEARS, A NEW CHALLENGE HAS INSPIRED ECLIPSE CHASERS: PHOTOGRAPH EARTHLIGHT ILLUMINATING THE MOON...

On 11 August 1999, at exactly 12:32 (Summer Time), the shadow of the Moon left French territory, racing across Central Europe, Asia, and finally disappearing in the Bay of Bengal. Hundreds of millions of people across the Old World observed the finest spectacle in nature, while others, who had been waiting for this moment for decades, saw, with a lump in their throat, storm clouds cover the sky, and eclipse the eclipse... Among this innumerable crowd of men, women, and children, all with their eyes turned to the sky, and who were either lucky enough to witness the event or, unfortunately, missed it, there were budding eclipse chasers. For many, the passion for eclipses, or quite simply for astronomy, was born there and then, somewhere between Cornwall and Isfahan, during those '2 minutes 35 seconds of happiness' (as a song would have it). If they are stay-at-homes or do not have the chance to travel, they will have to be patient for a long time before seeing another total eclipse change day into night over Europe. In Spain, a fine total eclipse will occur on 12 August 2026. Subsequently, in France, two total eclipses will occur, separated by just nine years: 3 September 2081 and 23 September 2090. The youngest of our readers will doubtless see one or other, but, in the meantime, everyone else will have to travel the world to attend the next encounter between the Moon and the Sun.

And they will have plenty of chances to see partial, annular, and total eclipses of the Sun. There are no less than 79 between the years 2000 and 2035! Similarly, over the same period, the Sun and Earth will provide us with 82 partial or total lunar eclipses. Finally, astronomers will also be lucky enough, during the first century of the third millennium, to witness 14 transits of Mercury and, above all, on 8 June 2004 and 5 June 2012, two transits of Venus across the Sun. This last phenomenon is extremely rare: no one has seen one since the end of the 19th century.

Everyone who witnessed the total eclipse on 11 August 1999 will have understood, often too late, to what extent the phenomenon, which is simultaneously both new and ephemeral, is difficult to appreciate as a whole. Many photographers, for example, will have either seen a wonderful eclipse and failed to get any photos, or succeeded in their photography, but failed to see the eclipse, failed to achieve either, or, and far more rarely, succeeded both with their observations and their photographs. The next total eclipses will be no exceptions to the rule. As always, they will come as a surprise, with their suddenness and their intensity, to both novices and hardened eclipse chasers alike. Nevertheless, it is still possible to give some

■ End of an eclipse for Fred Espenak, the world's greatest specialist in the field. On 16 February 1999, the American astronomer was in Australia to photograph an exceptionally narrow annular eclipse of the Sun.

■ On 10 May 1994, a beautiful annular eclipse was observable from New Mexico. At maximum eclipse, 94% of the Sun was hidden.

advice to anyone who wants to seek out forthcoming eclipses. Initially, it is useful to divide observation of eclipses of the Sun into two major, distinct categories. One group covers partial and annular eclipses, as well as the partial phases of total eclipses; and the other includes just totality among total eclipses.

In effect, in the first case, all that is involved is watching, with plenty of precautions, the Sun as it is more or less partly covered, but never completely hidden, by the Moon. The spectacle occurs in full daylight. In the second case, on the other hand, even if the total eclipse occurs in the middle of the day, basically one is observing a night-time celestial phenomenon.

During the eclipse of 11 August 1999, it was hammered home that observing the Sun without protecting the eyes is extremely dangerous, and may cause severe, temporary or permanent, damage to the retina, and thus to sight. Our star is, in fact, so bright that even when more than 99% is hidden by the Moon, it is still too bright for the eyes. To put it another way, both the partial phases of solar eclipses and annular eclipses must always be viewed with suitable protection for the eyes. Special glasses of Mylar or, even better, of black polymer for naked-eye observation, filters of special glass for refractors, reflectors, and telephoto lenses. All are available from optical firms that specialize in astronomical equipment. Such filters reduce the brightness of the Sun to a reasonable level, decreasing it by a factor of around 100 000.

■ Waiting. For eclipse chasers, two or three years pass between 'totals'. But the way in which time accelerates when that long-awaited instant arrives is unbelievable.

Partial eclipses – there will soon be several fine ones over Europe, on 31 May 2003, 3 October 2005, 29 March 2006, and 1 August 2008, and three over North America, on 14 December 2001, 10 June 2002, and 8 April 2005 – obviously do not have the power – and do not provoke the same emotions – as a true 'total'. Nevertheless, the whole of Europe looked up into the sky to admire the eclipse of 11 August 1999, which, outside the narrow band of totality, was naturally a partial eclipse. From which we may bet that the next 'partials' will be widely followed. Seeing the black disk of the Moon start, at the time predicted, to eat away at the disk of the Sun remains a source of fascination, especially when the phenomenon can be followed through the eyepiece of a telescope. Then, numerous details, which the naked eye – protected, let us emphasize for the last time, by special glasses – cannot detect, may be appreciated. First, the dark disk,

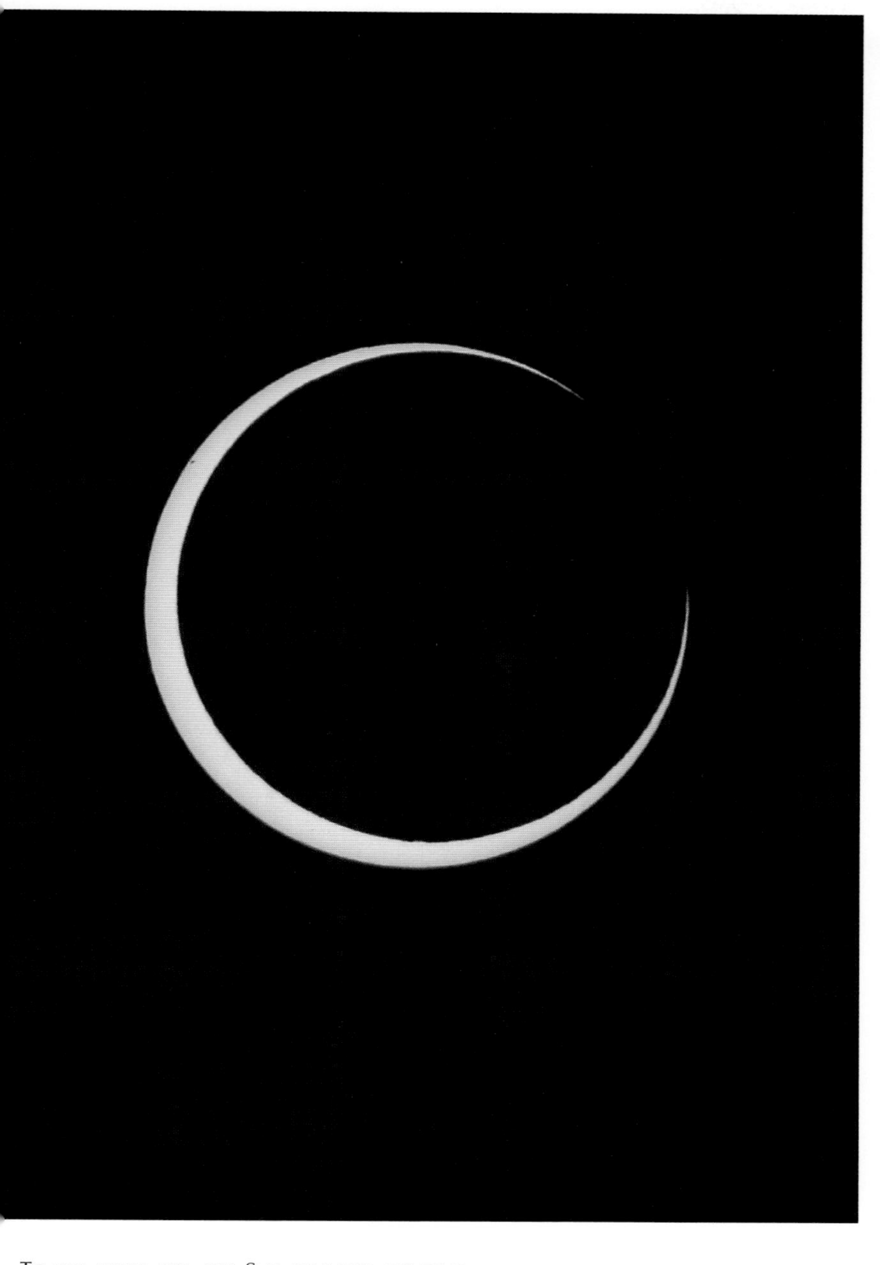

To the naked eye, the Sun remained blinding: these images required the use of a powerful filter.

which is far from circular, exhibits all the subtleties of relief: mountains and valleys may be seen around the lunar limb. Then the Sun itself may have some surprises: if sunspots dot its surface, they may be seen to disappear slowly behind the Moon. The luckiest ones may even be able to observe the eclipse with a coronograph, which enables one to see the chromosphere and solar prominences by means of a sophisticated system of optics, which produces artificial eclipses. Apart from professional observatories, there are nowadays small amateur coronographs, at a relatively affordable cost, that may be adapted for use with an astronomical refractor of 80–150 mm in diameter. Observing a partial eclipse with such an instrument is an unforgettable experience. One can, in fact, see the Moon approach the Sun, as a black disk silhouetted against the bright sky, before the eclipse proper begins. This is something that it is quite impossible to see without such an instrument.

Annular eclipses, although far less striking than totals, are still celestial phenomena that eclipse chasers do not hesitate to go to the ends of the Earth to observe, as with total eclipses. First, because no

■ Photography of solar prominences and details of the corona requires significant equipment. Here, a telescope 200 mm in diameter is being used with a long telephoto lens.

annular eclipse resembles any other. It is, as we have seen in Chapter 4, because the diameter of the Moon is insufficient to cover the Sun completely, that these particular eclipses arise, But the ratio of the diameters of the Sun and the Moon is never the same. In certain cases, as on 10 May 1994, at maximum eclipse, the Moon leaves a large ring of light around it. In other cases, such as that of 16 February 1999, the ring of sunlight is extraordinarily narrow. Finally, as on 30 May 1984, the eclipse appears to hover between being total and annular. For an instant, the ring of sunlight breaks up, flickers with light, and may even enable one to glimpse the lower corona in a darkened sky, in which planets are visible. The sight is one that has some of the force of a total eclipse, although it can never be as powerful.

And then there is, as many of our readers already know, the most beautiful sight in nature – a total eclipse. This may be divided into two distinct stages. First, the partial phase, which takes an hour and a half before totality, and then, of course, the same amount of time after it. For most eclipse chasers, the partial phase is no more than a preparation for the 1, 2, 3, or 4 minutes of happiness that are to come, when the Sun will be

■ The start of an eclipse over Sacramento Peak Observatory, in New Mexico. This image is the result of a double exposure. First the Sun was photographed through a filter, then a second image of the solar tower was recorded on the same negative.

completely hidden by the Moon. For photographers, it is also the time to make one last-minute check of the equipment, to frame the Sun against the landscape where it will be photographed, and so on. As with a true partial eclipse, the partial phase of a total eclipse should be followed with all the necessary precautions until the moment when the Sun is completely hidden. A total eclipse of the Sun is an event that has several different aspects, and may therefore be observed with very varied instruments. By all accounts, this most beautiful of all celestial phenomena, which engulfs the whole of nature for a few minutes, is particularly striking when viewed in the most natural manner: with the naked eye. This is the only way to take in the whole landscape suddenly plunged into night, to see here and there planets and stars, to admire the solar corona, notice distant parts of the horizon still illuminated by the Sun, and then watch the incredible sight of the Moon's shadow racing away towards the east.

Nevertheless, we must also advise any new eclipse chasers to take a pair of binoculars on any future expeditions. These, if they are of good quality, not too powerful, and are light, can provide an absolutely marvellous view of the eclipsed Sun. If they magnify by 7 to 10 times as a maximum, and have been carefully adjusted for the observer's sight – for example, by examining a star field on the previous night – and, if possible, set up on a photographic tripod to give the maximum visual comfort, they enable one to appreciate the subtleties of the plumes of the corona, to see in detail, as the naked eye cannot, the prominences that rise from the chromosphere, and also the ashen light on the Moon. Because a total eclipse lasts for all-to-brief a moment, and there is far more to see than just the Sun and the Moon, binoculars should be used for no more than twenty or thirty seconds once the total phase has started. It is essential to leave time to take in the scene with the naked eye. Binoculars have so many advantages for observing eclipses because they simultaneously have a wide field of view and good resolution, i.e., enable one to see fine detail. The solar corona, which is, as we may note in passing, after the Milky Way, the most extended object observed astronomically, may be seen in its entirety in a pair of binoculars, and in a very clear sky, one may even be able to see it more than 2° away from the disk of the eclipsed Sun.

■ Simplicity of use is one of the secrets of good photographs of eclipses. Here, a 600-mm telephoto lens is being used on a very stable tripod.

By comparison, the field of view of an astronomical refractor, or of a reflector, which an amateur astronomer might be tempted to use, is too small: 1° at the most. For this reason, observing an eclipse through too powerful an instrument is disappointing: the centre of the eyepiece field is filled with the dark disk of the Moon, and only the chromosphere, its prominences, and the lower corona are visible. Once again, binoculars, and the comfort that comes with binocular vision, are far preferable to a telescope.

Photographing eclipses

There are two profoundly different ways of approaching the most beautiful of all nature's spectacles: as an astronomer, or as a photographer. The former tries to represent the astronomical phenomena in images: the advance of the Moon over the disk of the Sun, and the different aspects of the solar atmosphere during the total phase – the thin thread of the chromosphere, the prominences, and, of course, the extensions to the corona. The latter tries to capture the eclipse's ambience, to represent it in a geographical or cultural context. In short, it is rather the terrestrial point of view that is the attraction.

Obviously, these two radically different photographic approaches require the use of completely different types of equipment. It even goes so far as to the choice of observing site, which may be influenced according to whether one opts for one or the other! For the astrophotographer, the choice is relatively simple. Whether they are a professional astronomer or simply a dedicated amateur, it is only a question of observing and photographing the eclipse under the best possible conditions. The only three criteria to be taken into account are the duration of the eclipse right along the path of totality; the meteorological statistics, again along this same line; and, of course, geographical accessibility. For an astronomer, the ideal would be to have a site in the exact centre of the zone of totality, where the eclipse would last the longest; that it should occur at midday; and that the Sun should be as high above the horizon, and as close to the zenith as possible. Given that

SOLAR ECLIPSES FOR PHOTOGRAPHERS

Aperture	2.8	4	5.6	8	11
50 ISO film	1/2 s	1 s	2 s	4 s	8 s
100 ISO film	1/4 s	1/2 s	1 s	2 s	4 s
200 ISO film	1/8 s	1/4 s	1/2 s	1 s	2 s

To photograph a landscape under the dim lighting of a total lunar eclipse, all that is required is to use a standard camera back, and a lens with a short focal length and wide aperture. The camera should be set to manual mode. This table gives an indication of the exposure times to use, with 100 ISO film and various aperture values, to photograph the landscape during the total phase.

SOLAR ECLIPSES FOR ASTRONOMERS

Aperture	5.6	8	11	16	22
Chromosphere	1/4000 s	1/2000 s	1/1000 s	1/500 s	1/250 s
Prominences	1/2000 s	1/1000 s	1/500 s	1/250 s	1/125 s
Lower corona	1/250 s	1/125 s	1/60 s	1/30 s	1/15 s
Middle corona	1/4 s	1/2 s	1 s	2 s	4 s
Outer corona	1/2 s	1 s	2 s	4 s	8 s

This table gives typical exposure times, for 100-ISO film and different aperture ratios (f/D), for lenses or astronomical instruments, to be used when photographing the various atmospheric layers of the Sun: the chromosphere and the different layers of the corona.

70 per cent of the Earth is covered with water, most of the time these conditions cannot be met – except in the middle of the sea! Pitching and rolling of ships, even when extremely small, are absolutely prohibitive for the use of powerful optical equipment (which amplifies any movement), so astronomers always end up alongside everyone else: on one or other of the seven continents.

Choices for a photographer who wants to emphasize the aesthetic aspects of the eclipse are both more interesting, more complex, and sometimes, more onerous. The very thing that astronomers want – an eclipse as high as possible in the sky – is one thing the photographer does not want. It is effectively impossible to include a Sun that is near the zenith within a terrestrial landscape, without using an extreme wide-angle lens, which has the double disadvantage of distorting images of the surroundings, and causing the image of the eclipsed Sun itself to be absolutely minute. This is why a photographer generally ignores the duration of the eclipse, and even the weather predictions, to have the chance of photographing the eclipse above the chosen site, whether that be a shoreline, a mountain, a castle, palace, temple, or whatever.

The plan, for any forthcoming eclipse, is always the same: first of all determine where the total eclipse begins (at sunrise), and where it finishes (at sunset). Then find out from maps (see pages 150 to 169) what types of countryside, what towns, what cultures, what geological or historical features will be covered by the Moon's shadow. Then choose your site. Once there, carefully check all the various locations, to put the finishing touches to the composition, and to rigorously ensure how the image will be framed. To do this, the best way is to visit the location a few days in advance, at the same time of day at which the total eclipse will occur. First, because the Sun is in exactly the same place in the sky, it is easy to check the field and the framing of the various objects. Second, the lighting of the chosen foreground will be the same – a contre-jour shot – as at the time of the eclipse, so it is easier to determine how the desired picture can be achieved, and whether it is really a good idea, and if it is possible to achieve with the equipment at one's disposal. In certain cases, the position of the Sun in the sky, relative to the chosen subject – a range of mountains, for example – may require a long search, in a vehicle, over several kilometres. In other cases, where the subject – a historical monument, for example – is relatively close to the photo-grapher, suitable framing requires the use of one's imagination. There remains the most difficult point of all: anticipating what will really happen during the eclipse: will one be alone in observing the event from this superb spot? If it is a historic site, it is highly unlikely, and on the day of the eclipse, it is prudent to be in place several hours in advance.

■ IT IS POSSIBLE TO USE SEVERAL CAMERAS AT ONCE. HERE, FRED ESPENAK AWAITS AN ECLIPSE WITH TWO REFRACTORS AND A CAMCORDER FITTED TO AN EQUATORIAL MOUNTING.

Because an eclipse generally lasts between 1 and 4 minutes, it is essential to have one or two camera backs with automatic film advance. Astrophotographers mount them at the focus of a powerful telephoto lens, or even of a telescope; and photographers mount them, with their lens, on a very stable photographic tripod. It is also possible, as this author does, to try to combine the two types of photography. All the photographs in Chapter 1, and most of those shown here, for example, were obtained with two Nikon F4S camera backs, Nikon lenses of 20 mm, 35 mm, 58 mm, 180 mm, 600 mm, and 800 mm focal length, the last being used occasionally with a 1.4 x focal-length multiplier. Such a spread of focal lengths, with progressively smaller fields and greater resolution, enables one to cover all possible (and conceivable) configurations that may occur during a total eclipse (see the Table). If at all possible, it is advisable to use identical cameras, whose controls are well-known to the photographer. Flexible cable releases help when taking the photographs, by eliminating or minimizing vibrations (sometimes caused by excessive emotions when releasing the shutter). The stress during a total eclipse, especially if photographers are in the middle of a large crowd, should not be under-estimated. It often leads to errors in camera handling. The most frequent of these include forgetting to alter the exposure times or the diaphragm settings during the total phase; and taking too many images just before the eclipse, which generally results in the camera stopping on the last frame at the crucial moment of totality! One golden rule: simplicity, for an eclipse photographer, is one guarantee of success. What must be avoided at all cost are attempting to be too ambitious, tinkering around and making changes at the last minute. A total eclipse, because of the emotional shock that it causes, really does reduce the capacity of most observers to think and act clearly: a considerable number of photographers found this out to their cost during the eclipse of 11 August 1999.

And yet, despite appearances, and our warnings, a total eclipse is a fairly easy subject for photographs, which does not require extremely sophisticated equipment, special film, nor particularly tricky shots. In fact, all it really comes down to, as far as the photographer is concerned, it to obtain a good contre-jour picture. Such a photograph copes well with slight errors in exposure times or diaphragm settings. Similarly, for the astronomer using a telephoto lens or a telescope, an eclipse is quite accommodating: the enormous range in luminosity between the chromosphere and the corona, allows one to use almost any combination of exposure and f-stop, practically without the risk of going wrong!

In practice, we would advise using negative or reversal films with speeds of between 50 and 200 ISO. Such a speed, which retains sharpness and image quality, is suitable for most of the conditions involved. The tables on p.146 give the exposure times to be used, as a function of the most common aperture values, for the various subjects of interest in a total eclipse. The first table is for photographers who want to frame the eclipse in a terrestrial setting. The second, by contrast, is intended for astrophotographers, who, using lenses of 300 to 2000 mm in focal length, want to capture the image of the Sun and Moon in a small part of the sky.

In practice, for a photographer, once a diaphragm setting has been chosen – f/4 to f/5.6 seems a good compromise – all one needs to do is make several exposures, 'bracketing' the average value given. Using the camera in manual mode with, for example, a setting of f/5.6, it will suffice to cover a range of

FIELD SIZES AND ANGULAR SIZES

Lens focal length	Field covered by 24 x 36 mm frame	Size of Sun and Moon (average 30′)	Size of the corona (about 2°)
20 mm	60° x 90°	0.2 mm	0.8 mm
28 mm	50° x 73°	0.3 mm	1.2 mm
35 mm	38° x 55°	0.4 mm	1.6 mm
50 mm	27° x 40°	0.5 mm	2.0 mm
135 mm	10° x 15°	1.3 mm	5.0 mm
180 mm	7° x 11°	1.5 mm	6.0 mm
300 mm	4°30′ x 6°50′	2.8 mm	11.0 mm
400 mm	3°30′ x 5°	3.7 mm	15.0 mm
500 mm	2°45′ x 4°	4.6 mm	18.0 mm
600 mm	2°15′ x 3°25′	5.6 mm	22.0 mm
800 mm	1°45′ x 2°30′	7.4 mm	30.0 mm *
1000 mm	1°20′ x 2°	9.2 mm	37.0 mm *
1600 mm	50′ x 1°10′	14.8 mm	60.0 mm *
2000 mm	40′ x 1°	18.4 mm	74.0 mm *

This table enables one to see, at a glance, the photographic field given by most normal lenses, telephoto lenses, and commonly found telescopes, with the standard format of 24 x 36 mm. Next, we give the size, on the film, in millimetres, of the eclipsed Sun (and thus of the Moon as well) as a function of the focal length. Finally, photographers and astronomers often forget that the solar corona, in a very clear sky, may be followed out to 2° or even 3° from the Sun. The right-hand column shows the size, on the film, of an outer corona that is about 2° in diameter. An asterisk (*) indicates that lenses longer than 600 mm do not allow the whole of the outer corona to be included in a single frame.

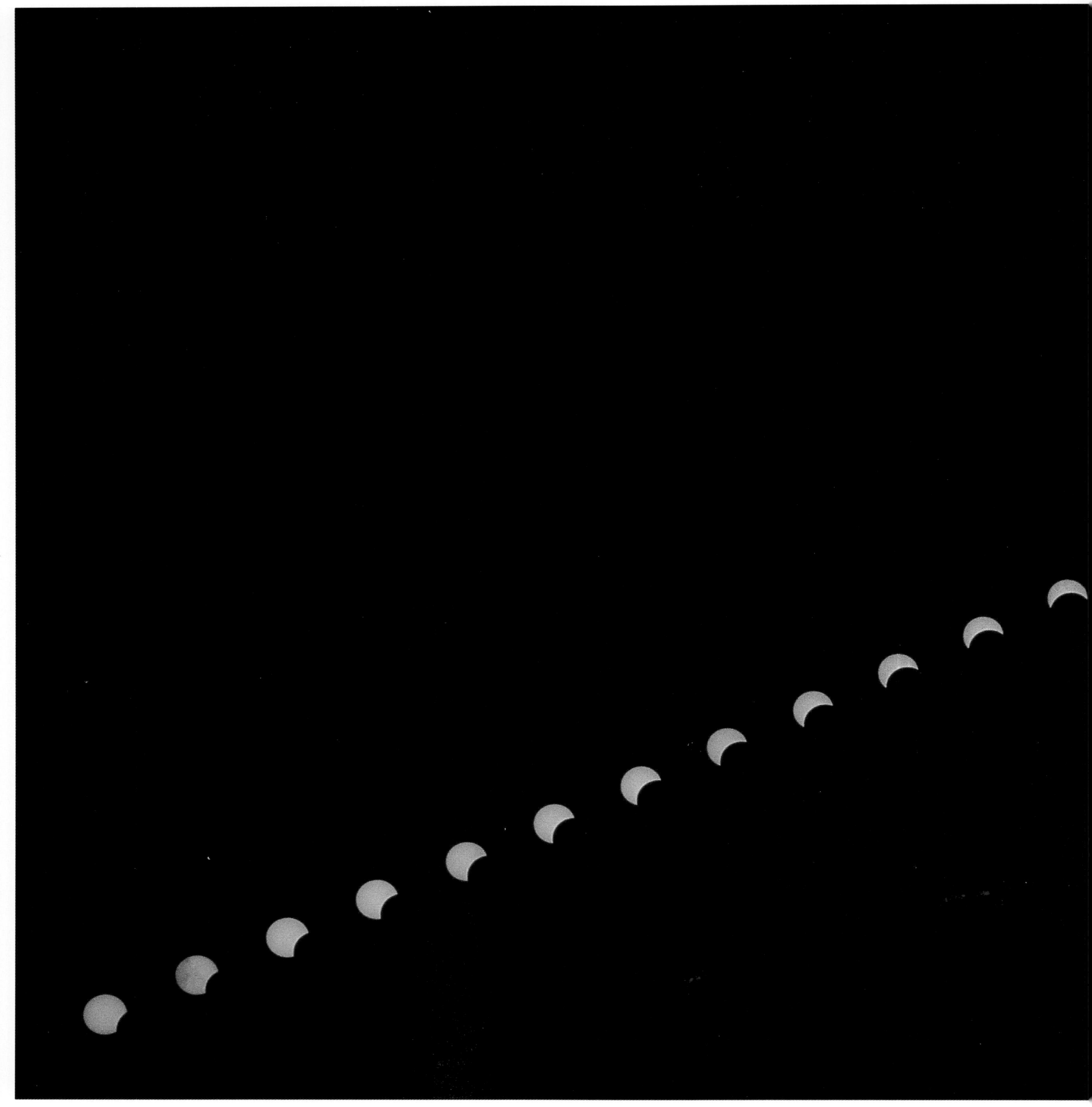

■ MULTIPLE EXPOSURES REQUIRE PERFECT WEATHER, GOOD PREPARATION, AND A LOT OF INNER CALM. FOR THIS PARTIAL ECLIPSE, AKIRA FUJII RECORDED THE PHASES

exposures each side of the value given, here 1 sec for 100 ISO. Such as: 1/4 s, 1/2 s, 1 s, and 2 s. By bracketing the exposure – as professionals call it – it is practically impossible to lose the picture. But why not simply restrict oneself to a single exposure as recommended by the table? Quite simply because no eclipse ever resembles the one before, and the amount of ambient light, during totality, is absolutely unpredictable. The luminosity of the corona, for example, may change by a factor of two from one eclipse to the next. In addition, the transparency of the sky, and the height of the Sun above the horizon may dramatically alter the conditions at the time of exposure.

Things are more complicated for astronomers, because, as the second table shows, it is obviously not feasible to photograph the whole of the eclipse, from the chromosphere to the outer corona with a single aperture setting. We would therefore advise astronomers who are fully conversant with their photographic equipment to alter the diaphragm setting and the exposure time, during the total phase. One might, for instance, start

OF THE EVENT WITH AN IMAGE EVERY THREE MINUTES, EXPOSED ON A SINGLE FRAME.

the eclipse with very short exposures and a small aperture (f/11, for example) to record the very bright chromosphere and the prominences. Then, a bit later, to alter the lens to a wider aperture (f/5.6, for example), and considerably extend the exposure times. Such a change seems like child's play when you read about it. Nevertheless, it may prove to be extremely difficult, during those all-too-short minutes of the eclipse and the strange night that it brings. ■

Total and annular eclipses of

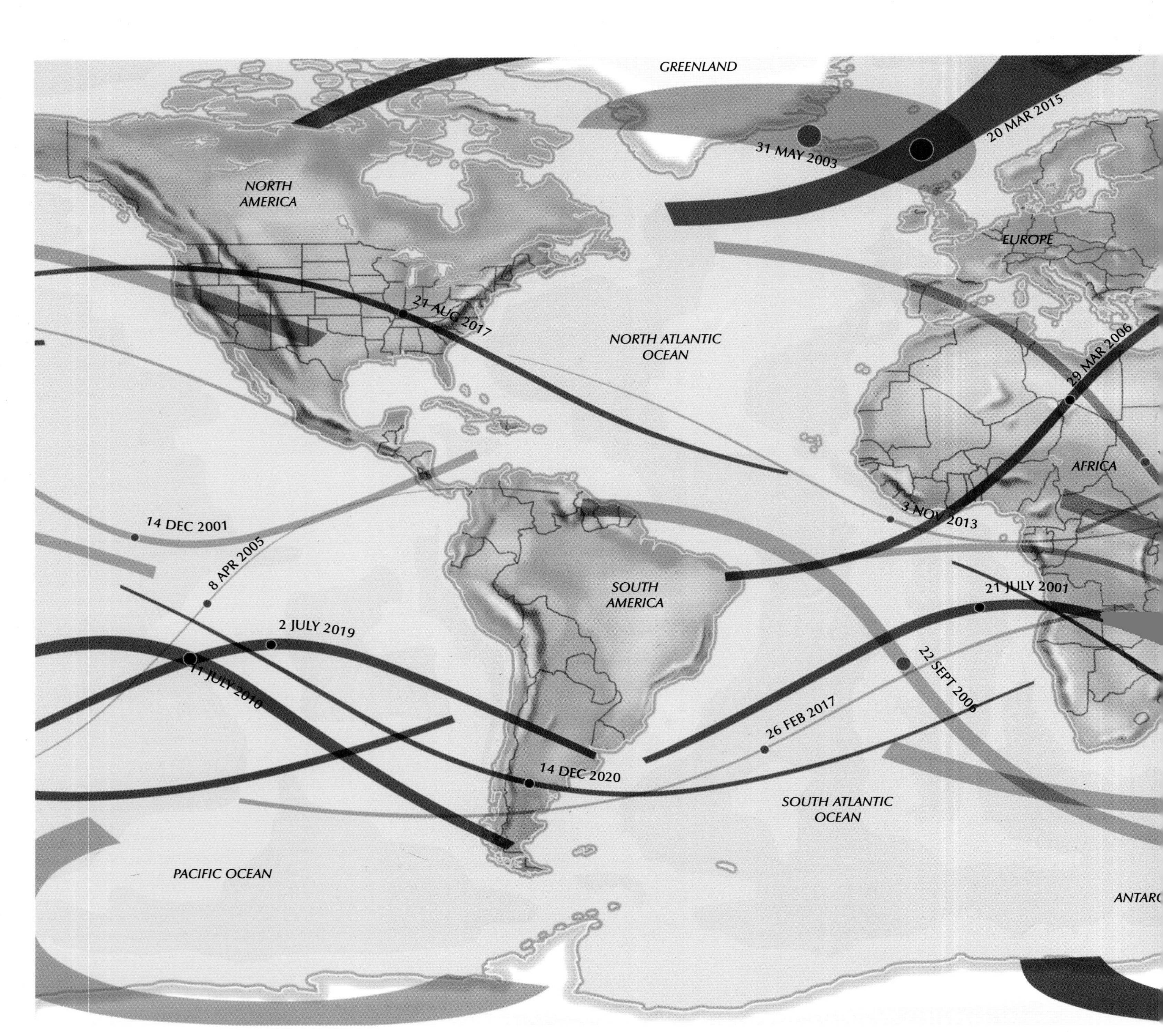

■ THIS MAP SHOWS ALL THE ANNULAR AND TOTAL ECLIPSES THAT WILL BE VISIBLE FROM ANY PART OF THE WORLD, BETWEEN 2001 AND 2020. EACH ECLIPSE, WHICH IS DATED, IS REPRESENTED BY A BAND, GREY FOR ANNULAR ECLIPSES, AND BLACK FOR TOTAL. EACH OF THESE BANDS REPRESENTS THE REGION WHERE AN ANNULAR OR TOTAL ECLIPSE WILL BE VISIBLE: THEY SIMPLY SHOW THE PATH OF THE MOON'S SHADOW ACROSS THE EARTH. THE PARTIAL PHASES, WHICH ARE VISIBLE OVER FAR LARGER AREAS, ARE NOT SHOWN. ECLIPSES START AT SUNRISE IN THE WEST, AND

the Sun between 2001 and 2020

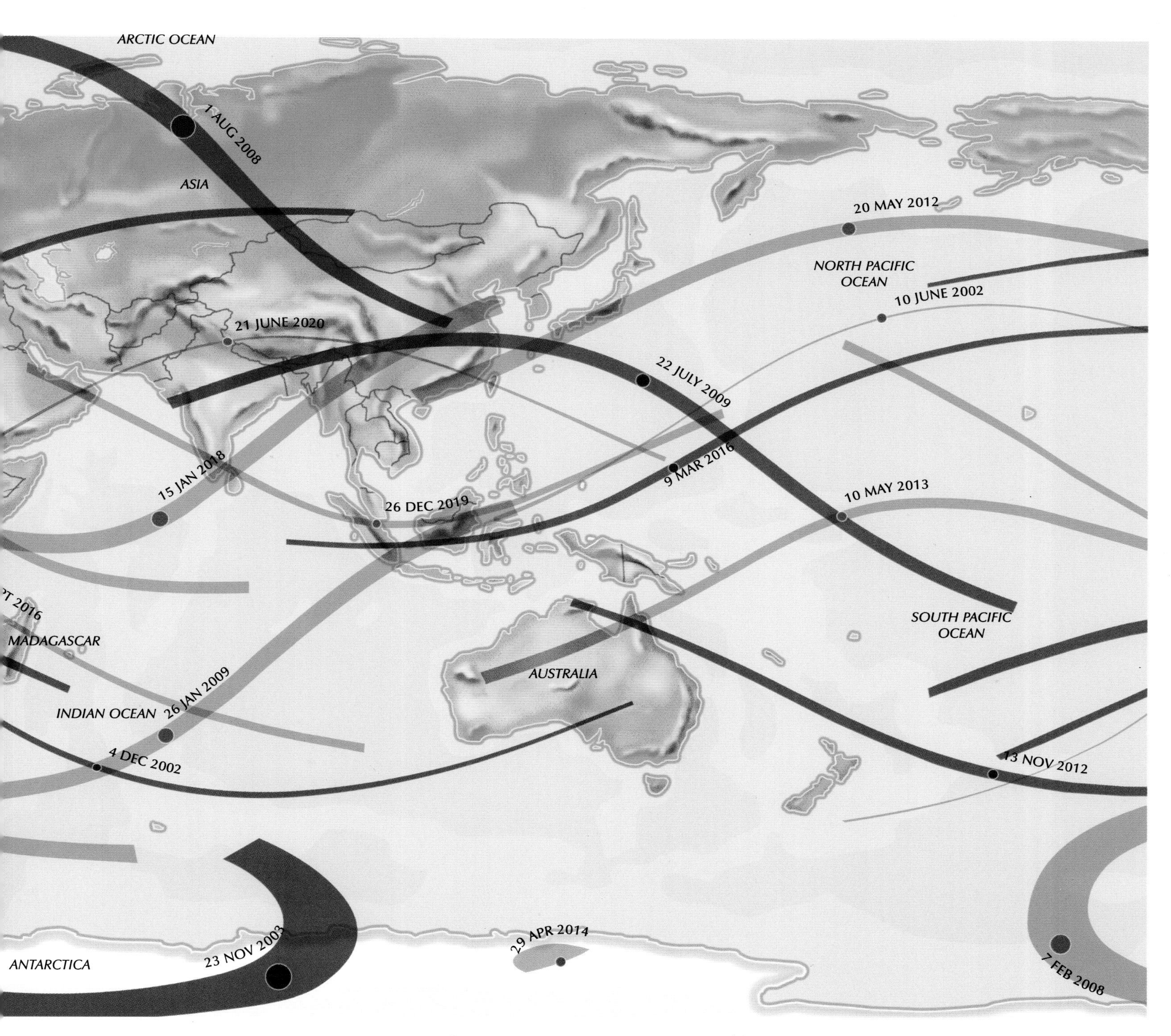

THUS ON THE LEFT ON THIS MAP, AND FINISH AT SUNSET, IN THE EAST. BETWEEN THOSE TIMES, THE SHADOW OF THE MOON SWEEPS OUT A REGION MORE THAN 10 000 KM IN LENGTH, BY BETWEEN TEN AND FIVE HUNDRED KILOMETRES WIDE. THE POINT WHERE MAXIMUM DURATION OF THE ECLIPSE OCCURS IS MARKED BY A SMALL DISK. EACH ECLIPSE IS GIVEN IN DETAIL, INCLUDING ITS PARTIAL PHASES AND THEIR DURATION, ON THE MORE DETAILED MAPS GIVEN ON PAGES 156 TO 169.

Total and annular eclipses of

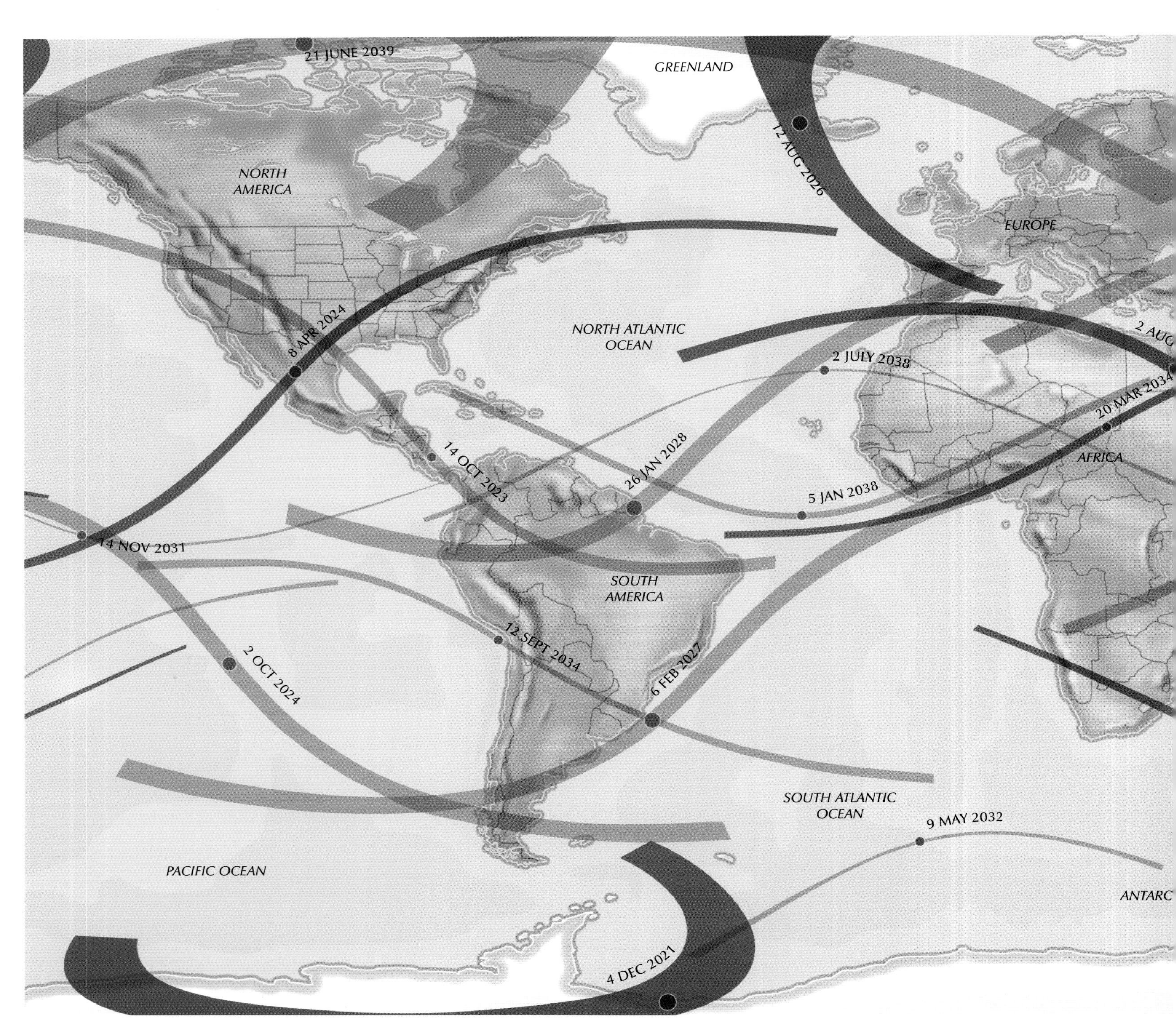

■ This map shows all the annular and total eclipses that will be visible from any part of the world, between 2021 and 2040. Each eclipse, which is dated, is represented by a band, grey for annular eclipses, and black for total. Each of these bands represents the region where an annular or total eclipse will be visible: they simply show the path of the Moon's shadow across the Earth. The partial phases, which are visible over far larger areas, are not shown. Eclipses start at sunrise in the west, and thus on the left on this map, and finish at sunset, in the east. Between those times, the shadow of the Moon sweeps out a

the Sun between 2021 and 2040

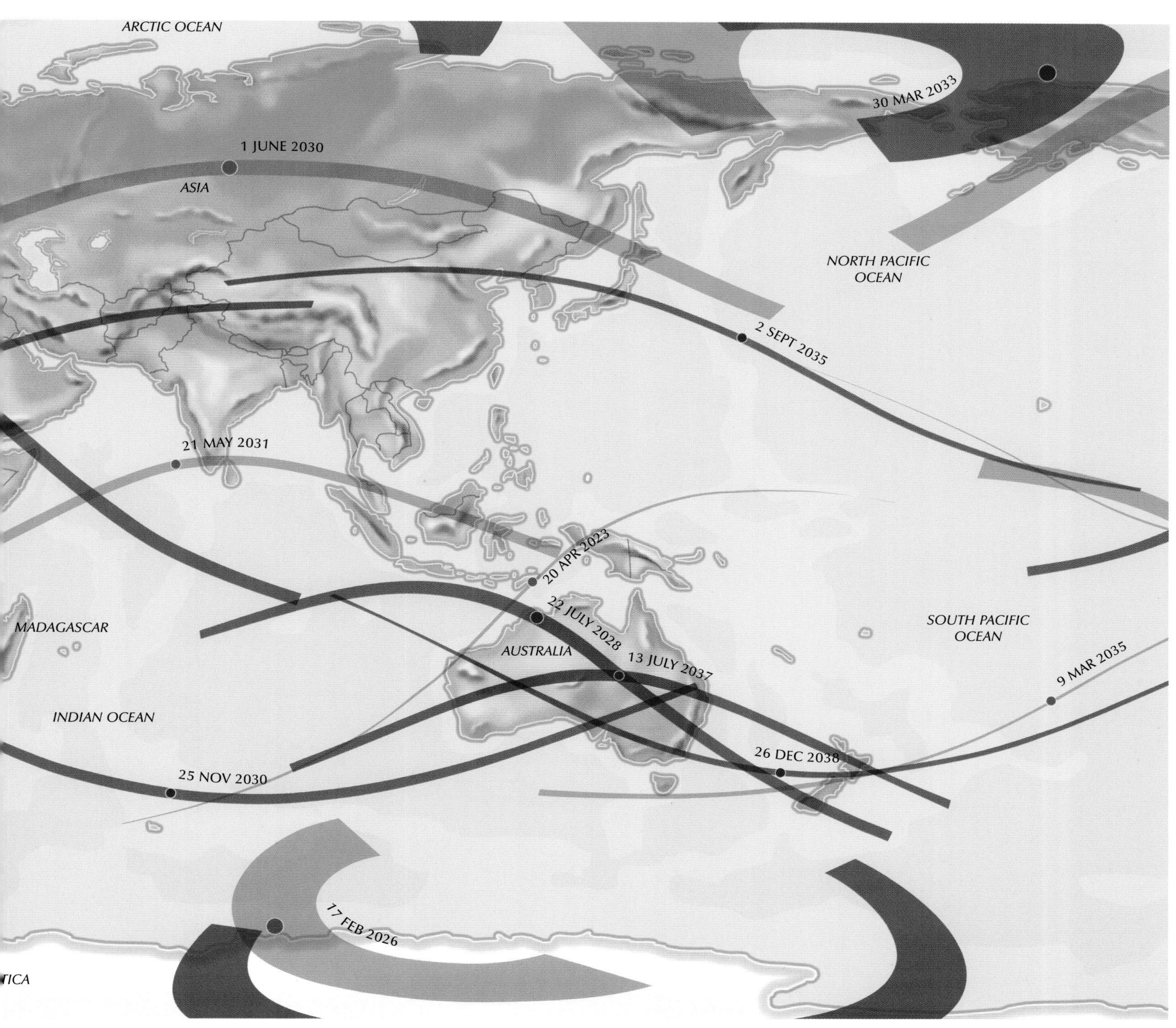

REGION MORE THAN 10 000 KM IN LENGTH, BY BETWEEN TEN AND FIVE HUNDRED KILOMETRES WIDE. THE POINT WHERE MAXIMUM DURATION OF THE ECLIPSE OCCURS IS MARKED BY A SMALL DISK. THE DURATION OF THE ECLIPSE DECREASES ON BOTH SIDES OF THIS POINT. IN EUROPE, THREE FINE ECLIPSES ENLIVEN THE THIRD DECADE OF THE CENTURY: 12 AUGUST 2026, A FINE TOTAL ECLIPSE FINISHES IN SPAIN AT SUNSET. ONE YEAR LATER, ON 2 AUGUST 2027, THE SOUTH OF SPAIN AND NORTH AFRICA ARE CROSSED BY A TOTAL ECLIPSE. FINALLY, ON 26 JANUARY 2028, SPAIN EXPERIENCES ITS THIRD ECLIPSE IN THREE YEARS!

Total and annular eclipses of

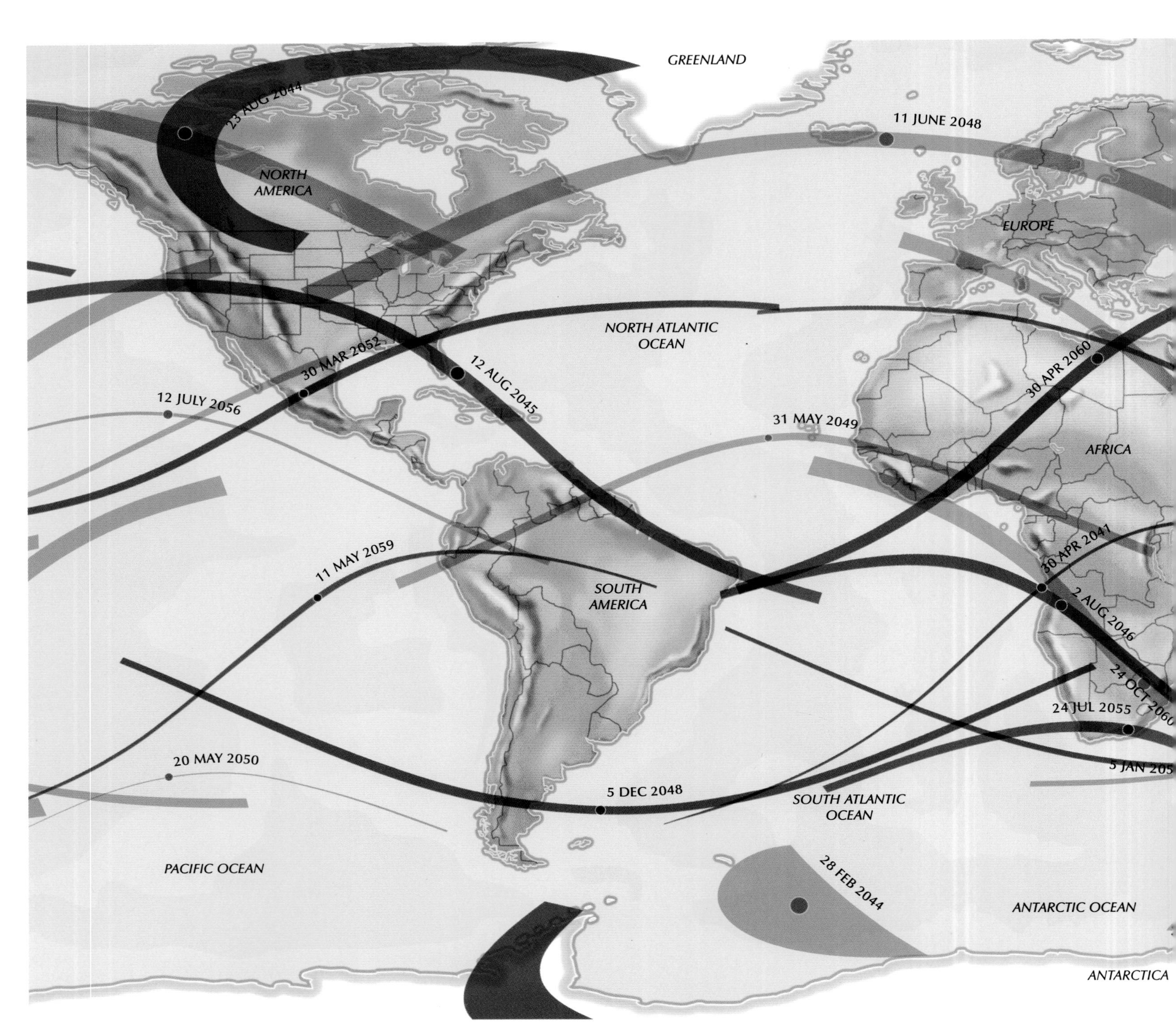

■ This map shows all the annular and total eclipses that will be visible from any part of the world, between 2001 and 2 Each eclipse, which is dated, is represented by a band, grey for annular eclipses, and black for total. Each of these b represents the region where an annular or total eclipse will be visible: they simply show the path of the Moon's shadow ac the Earth. The partial phases, which are visible over far larger areas, are not shown. Eclipses start at sunrise in the west, thus on the left on this map, and finish at sunset, in the east. Between those times, the shadow of the Moon sweeps c

the Sun between 2041 and 2060

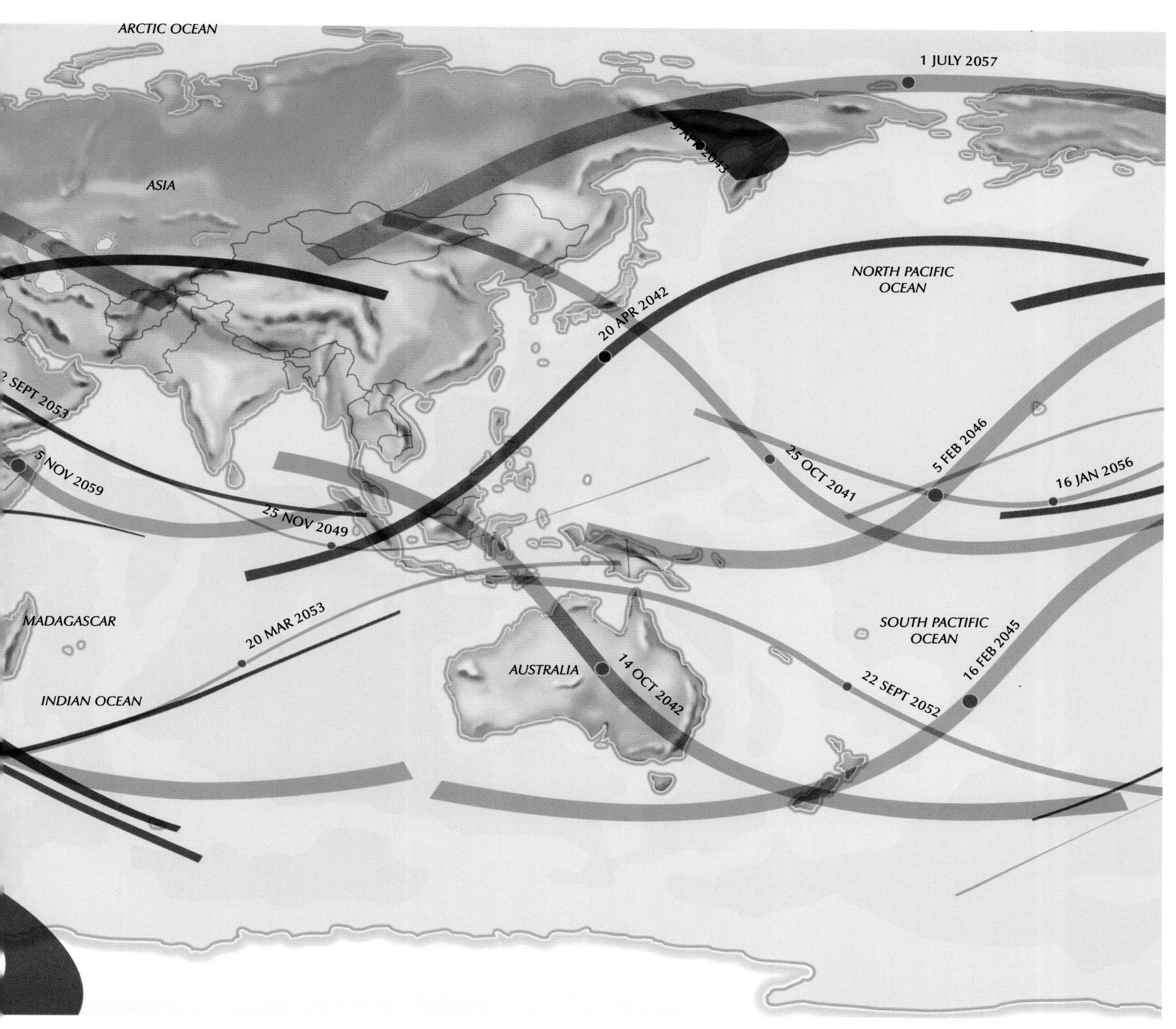

REGION MORE THAN 10 000 KM IN LENGTH, BY BETWEEN TEN AND FIVE HUNDRED KILOMETRES WIDE. THE POINT WHERE MAXIMUM DURATION OF THE ECLIPSE OCCURS IS MARKED BY A SMALL DISK. ECLIPSE CHASERS WILL ATTEMPT A FIRST: OBSERVING THREE CENTRAL ECLIPSES FROM EXACTLY THE SAME POINT! THE ECLIPSES OF 12 SEPTEMBER 2053, 5 NOVEMBER 2059, AND 30 APRIL 2060 WILL CROSS THE SAME REGION OF THE SAHARA, ON THE BORDERS OF LIBYA AND EGYPT

21 June 2001

Total eclipse
Duration: 4 min 56 sec

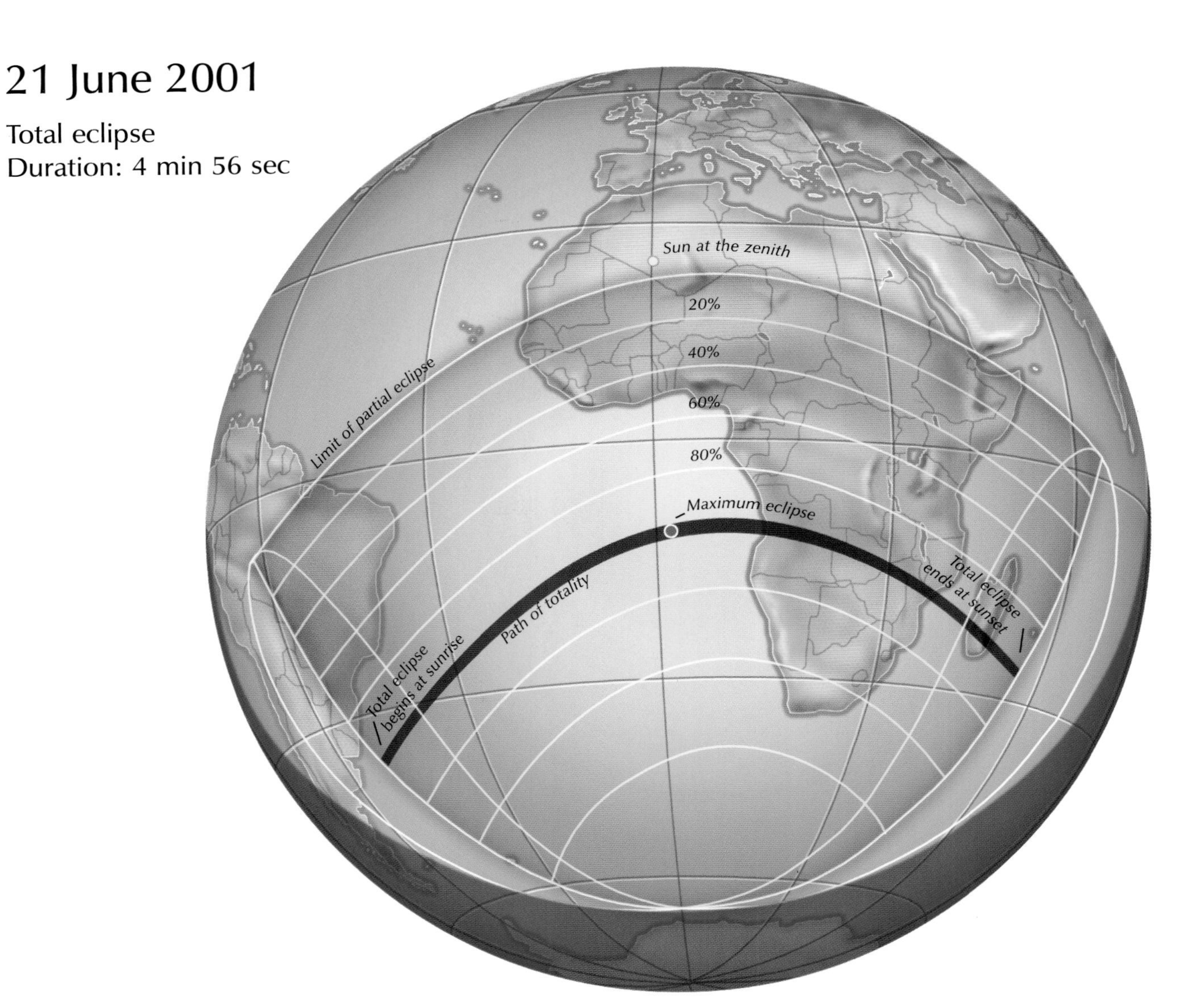

The following pages show all eclipses of the Sun, annular or total, visible to the end of the year 2020. The orientation of the maps corresponds to the view that one would have of our planet, as seen from the Moon, at the time of maximum eclipse. The small yellow disk on each map marks the point on the Earth where the Sun will be at the zenith at the time of maximum eclipse.

14 December 2001

Annular eclipse
Duration: 3 min 53 sec

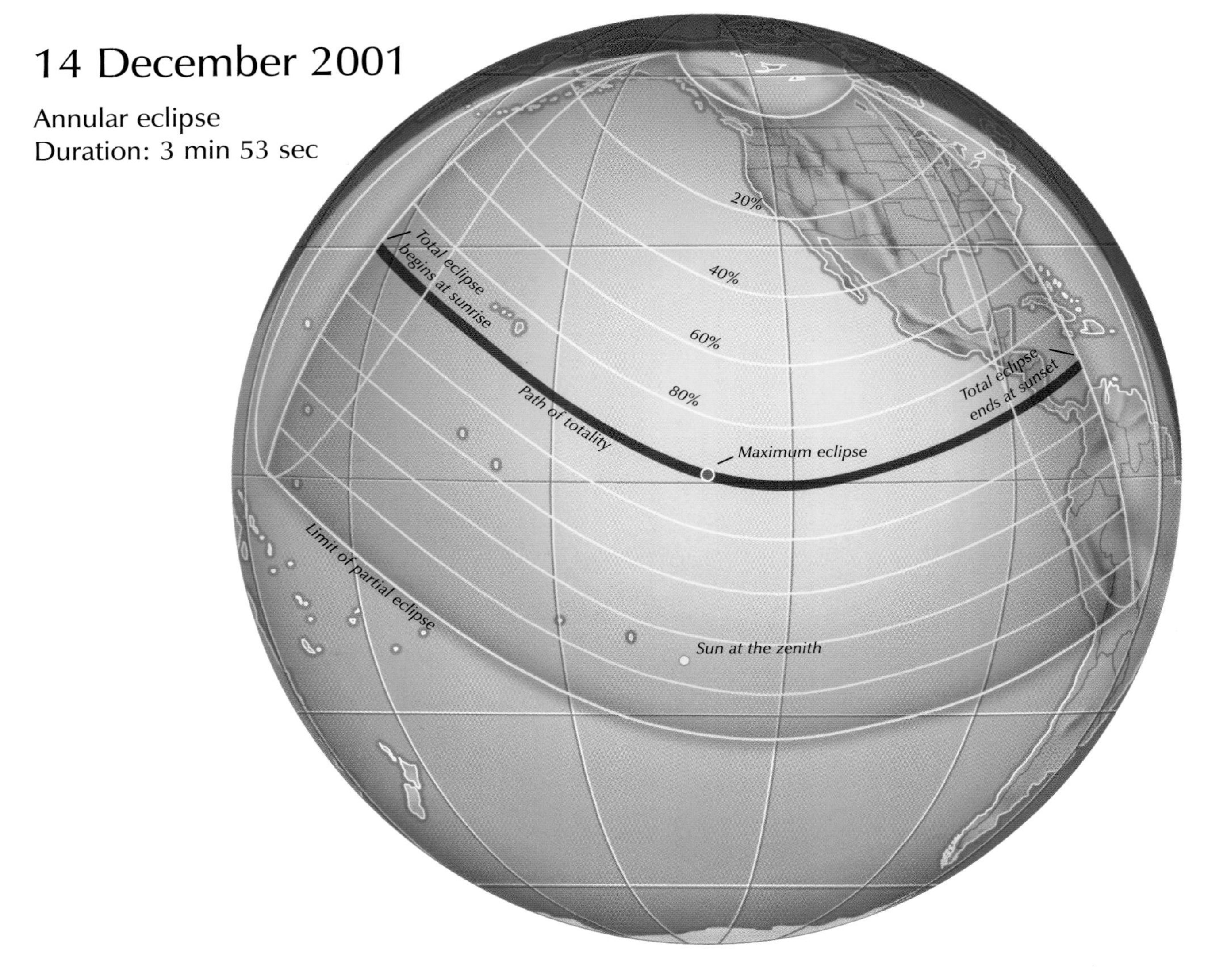

10 June 2002

Annular eclipse
Duration: 22 sec

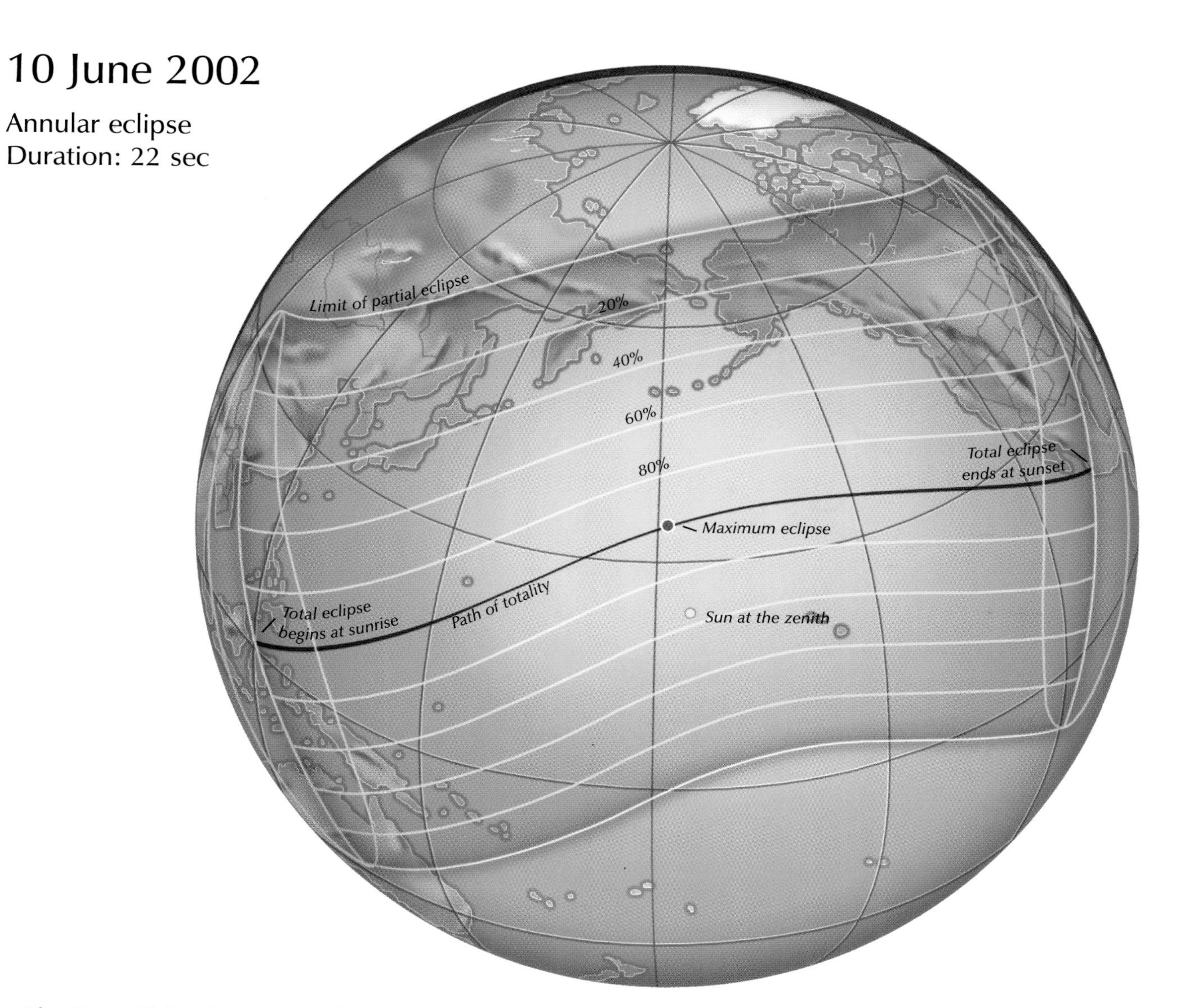

The Sun will lie close to the fine red supergiant Antares in Scorpius during the eclipse of 4 December 2002. The two bodies, just 5° from one another, will be capable of being photographed with a lens of 85–135 mm focal length. In Angola, this eclipse occurs just eighteen months after another total eclipse, that of 21 June 2001.

4 December 2002

Annular eclipse
Duration: 2 min 3 sec

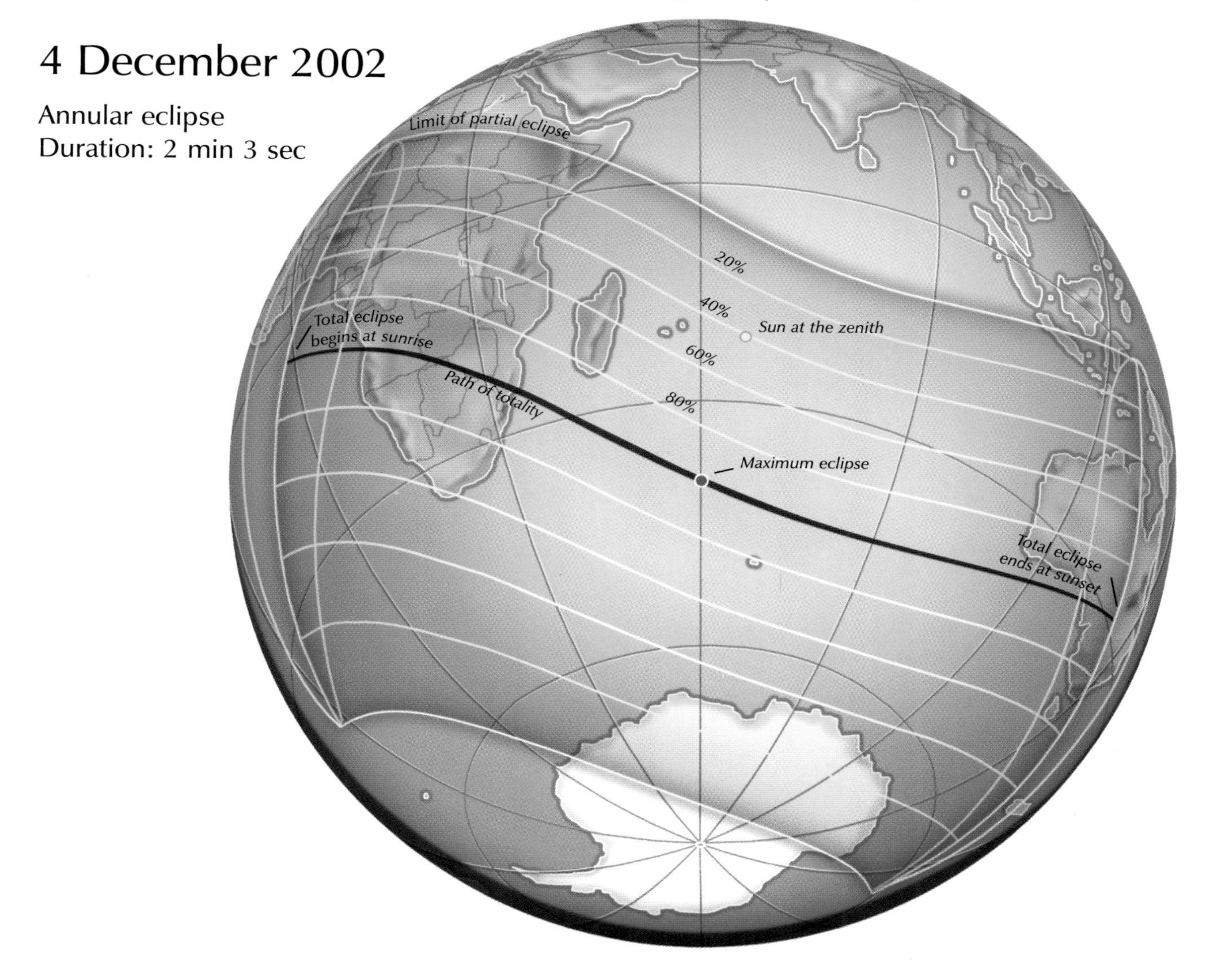

31 May 2003

Annular eclipse
Duration: 3 min 36 sec

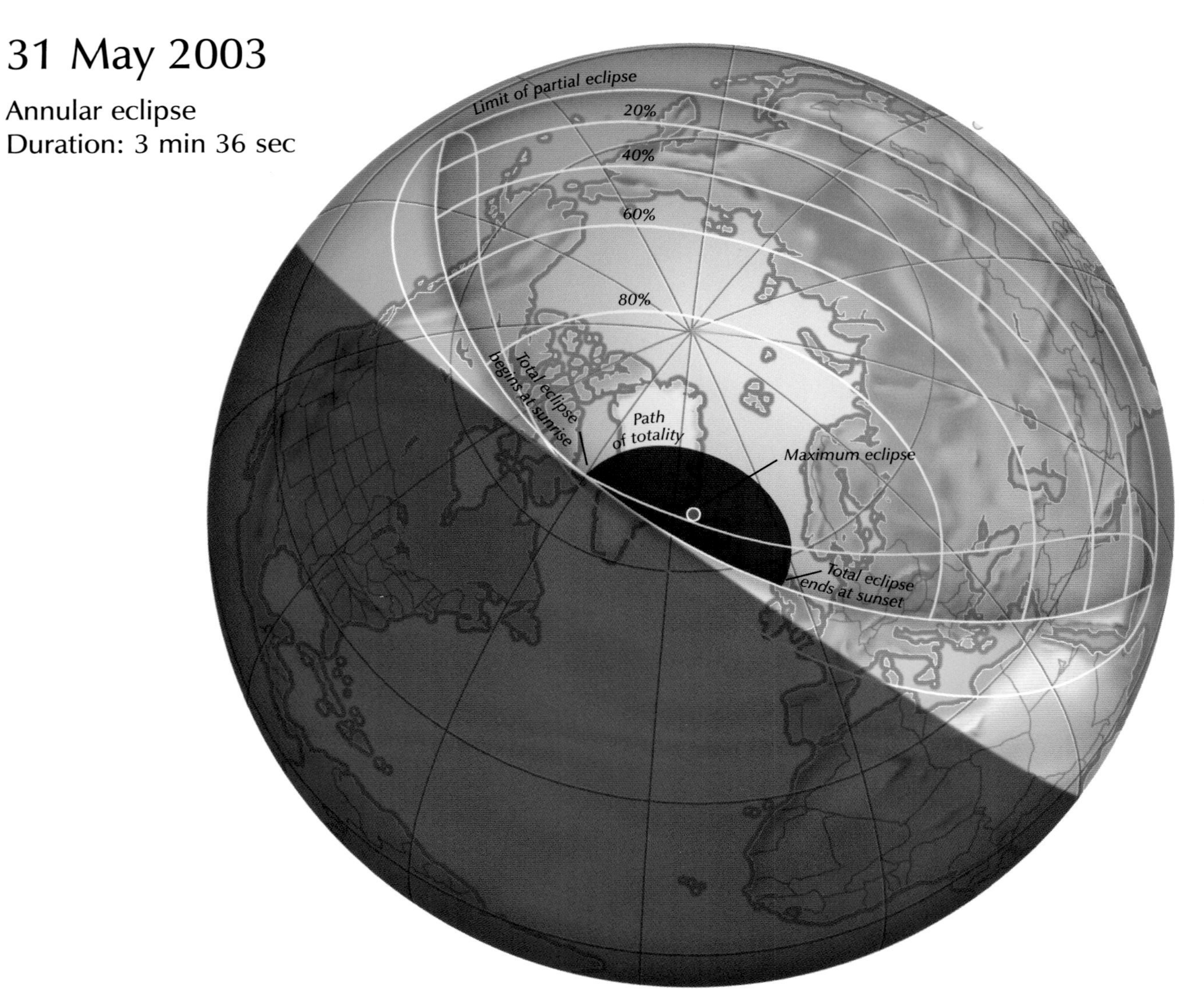

In 2003, eclipse chasers who want to indulge their passion, will need to travel, with a six-month interval, first to one pole and then to the other. On 31 May, the annular eclipse is visible from the north of Scotland to Greenland, whereas the total eclipse of 23 November is visible only from Antarctica, at the end of the southern spring, and a practically inaccessible region.

23 November 2003

Total eclipse
Duration: 1 min 57 sec

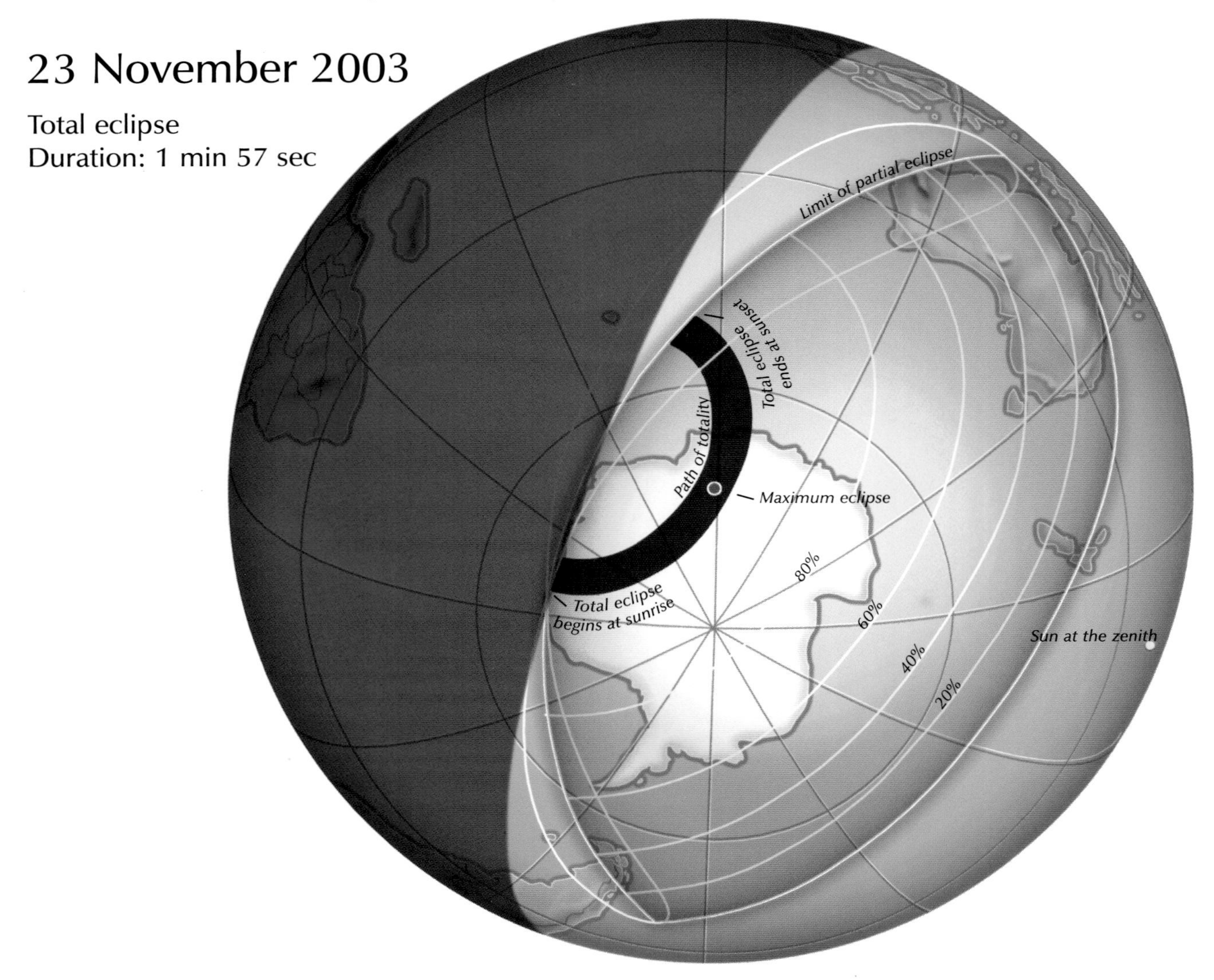

8 April 2005

Annular/total eclipse
Duration: 42 sec

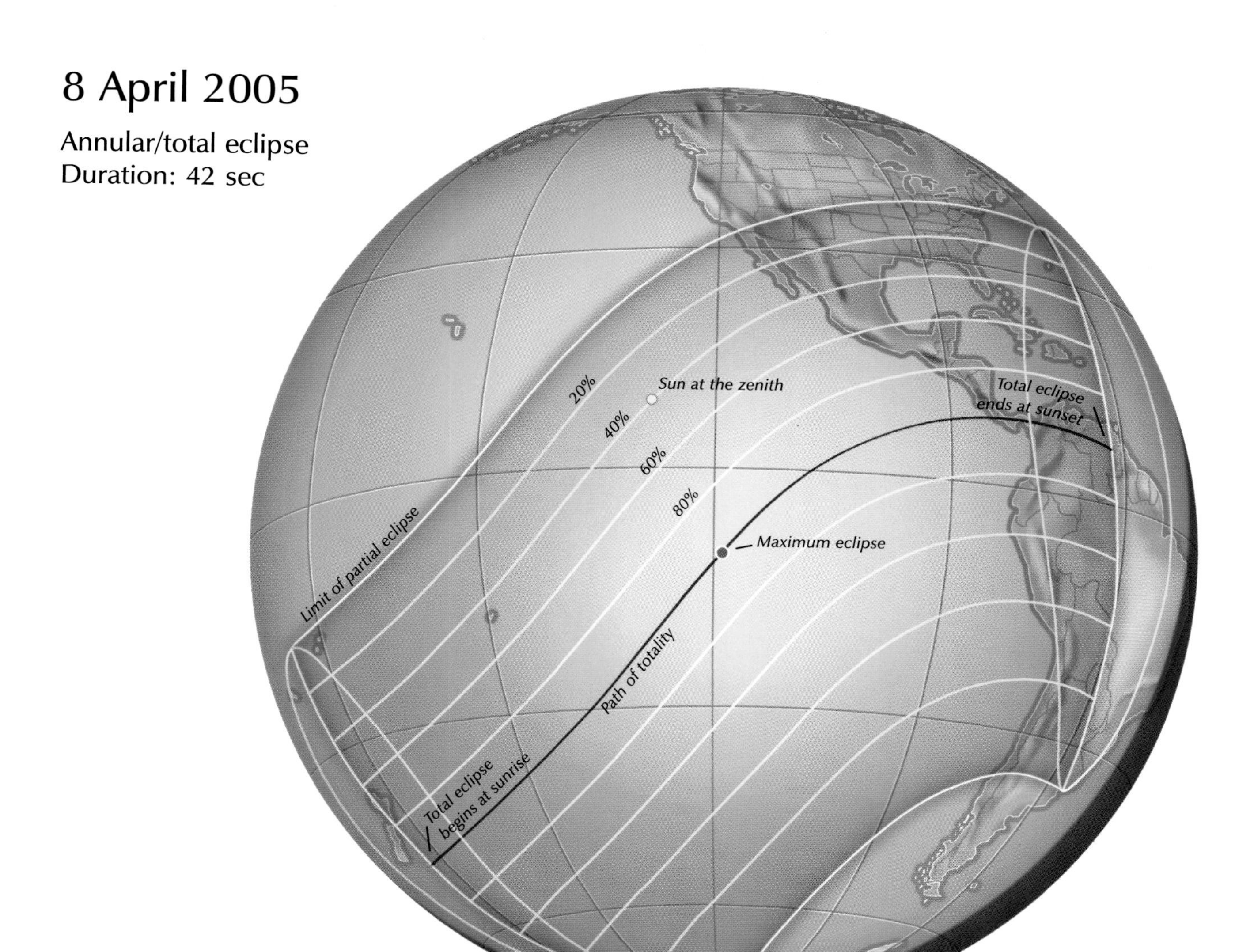

On 8 April 2005, the narrow pencil of lunar shadow scarcely touches the Earth. The eclipse, with an exceptionally short duration – between a few seconds and 42 seconds at maximum – is total only in the middle of the Pacific. But when the shadow of the Moon reaches the coast of Central America, the eclipse, lasting just 5 seconds, is just annular.

3 October 2005

Annular eclipse
Duration: 4 min 3 sec

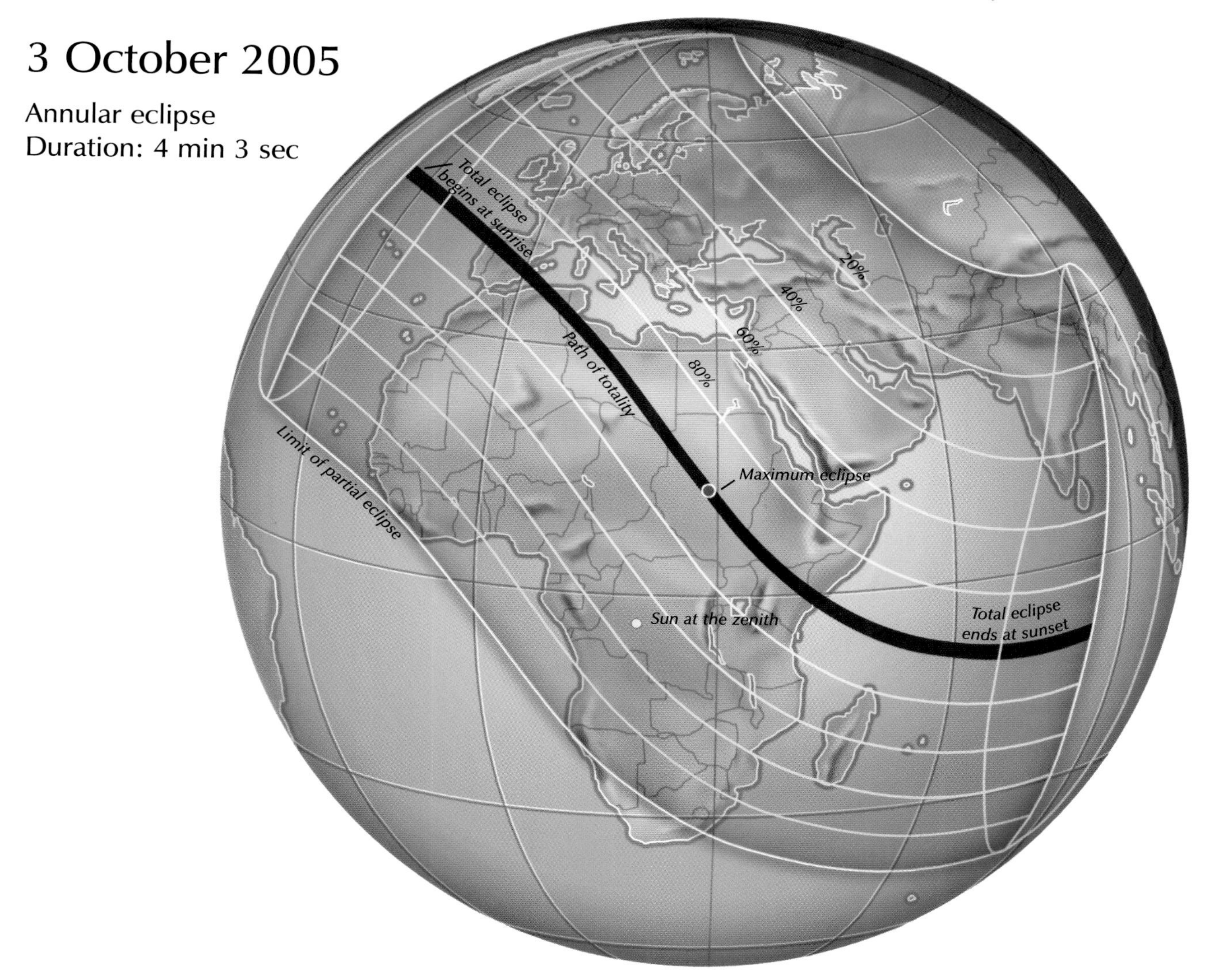

29 March 2006

Total eclipse
Duration: 4 min 6 sec

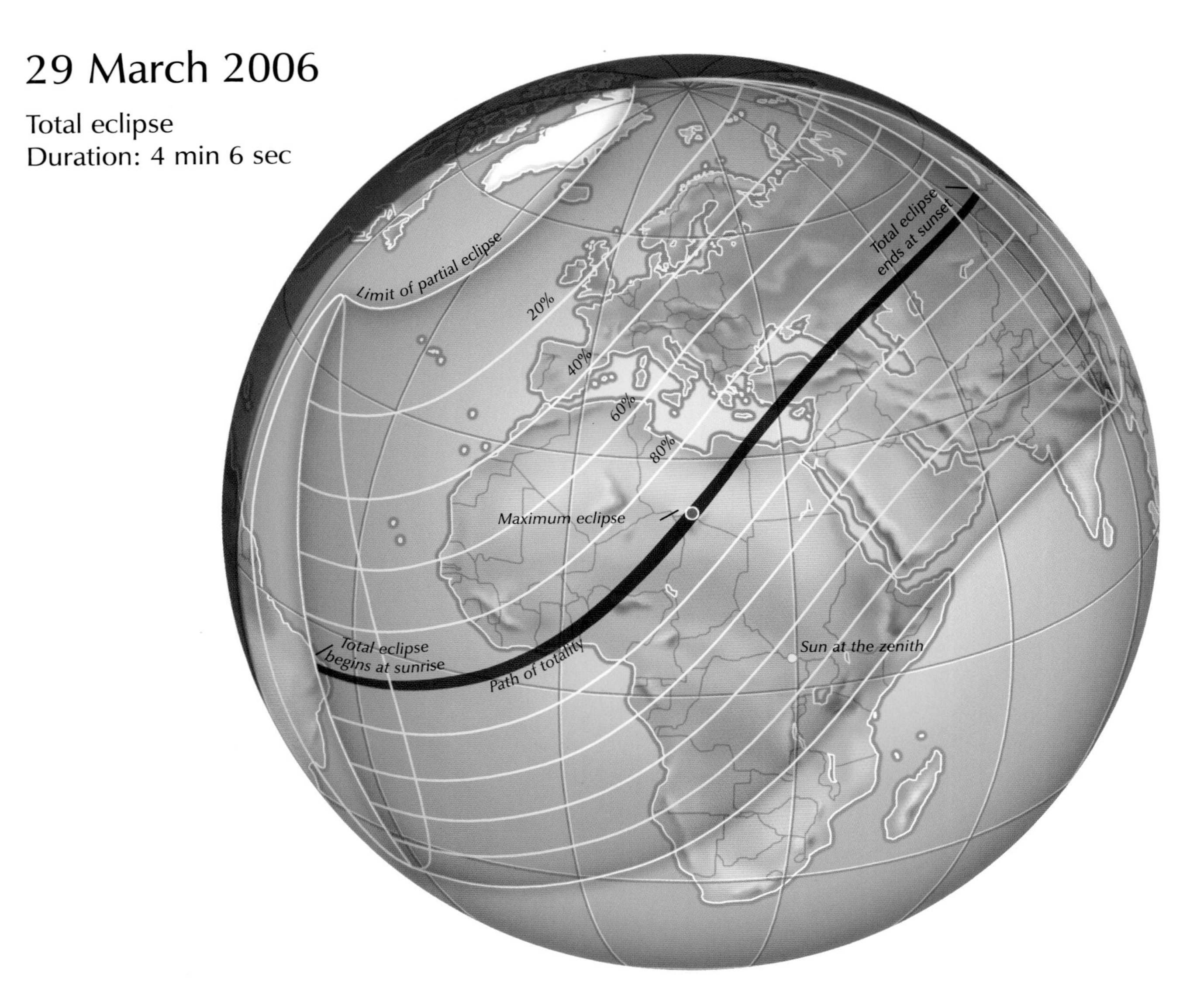

The eclipse of 29 March 2006 crosses a large part of the African continent, with a high probability of fine weather, in particular in Niger, Libya, and Chad. In contrast, the annular eclipse of 22 September 2006, visible in the deserted expanses of the Atlantic Ocean, is accessible, at sunrise, only in Surinam and French Guyana.

22 September 2006

Annular eclipse
Duration: 7 min 9 sec

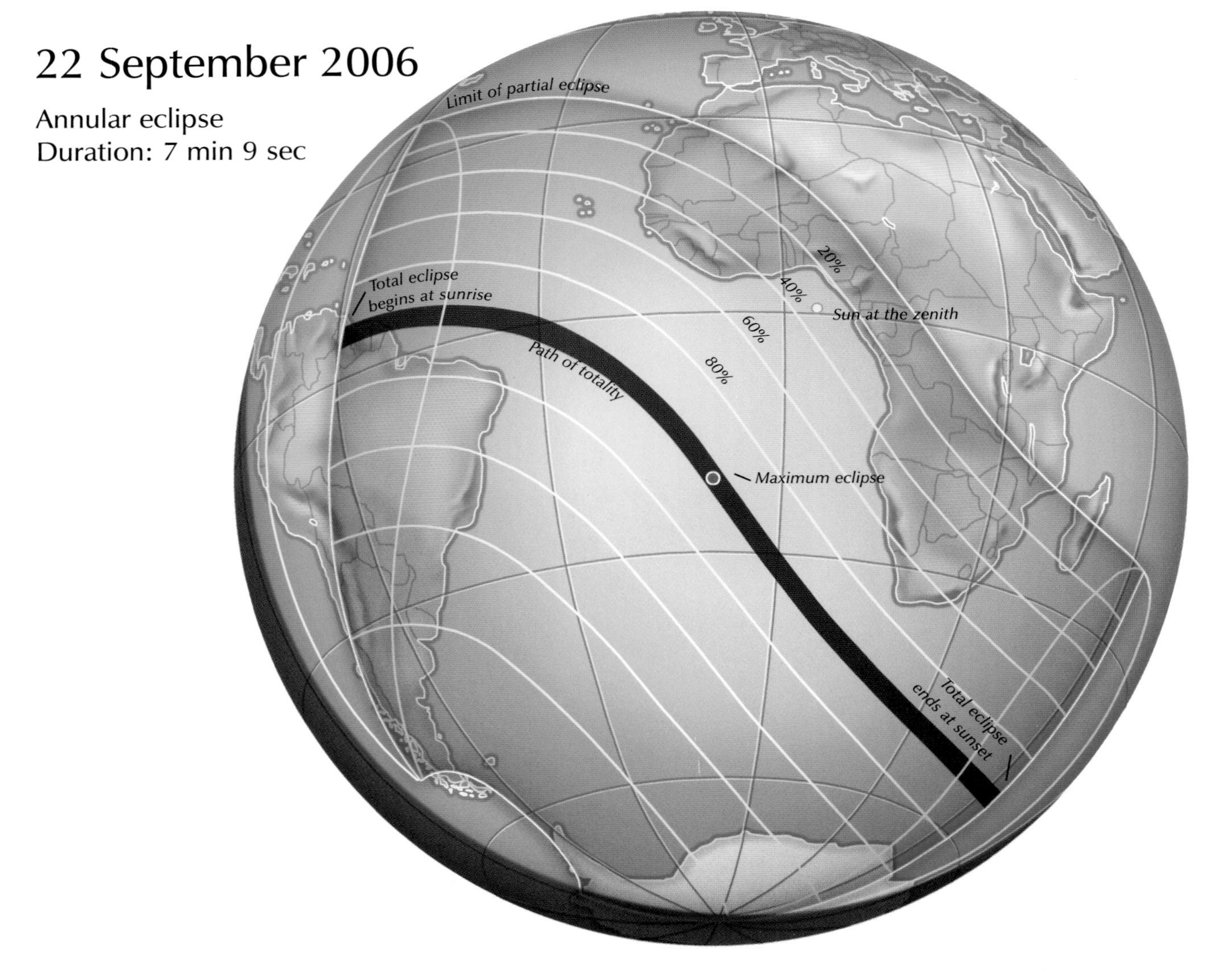

1 August 2008

Total eclipse
Duration: 2 min 27 sec

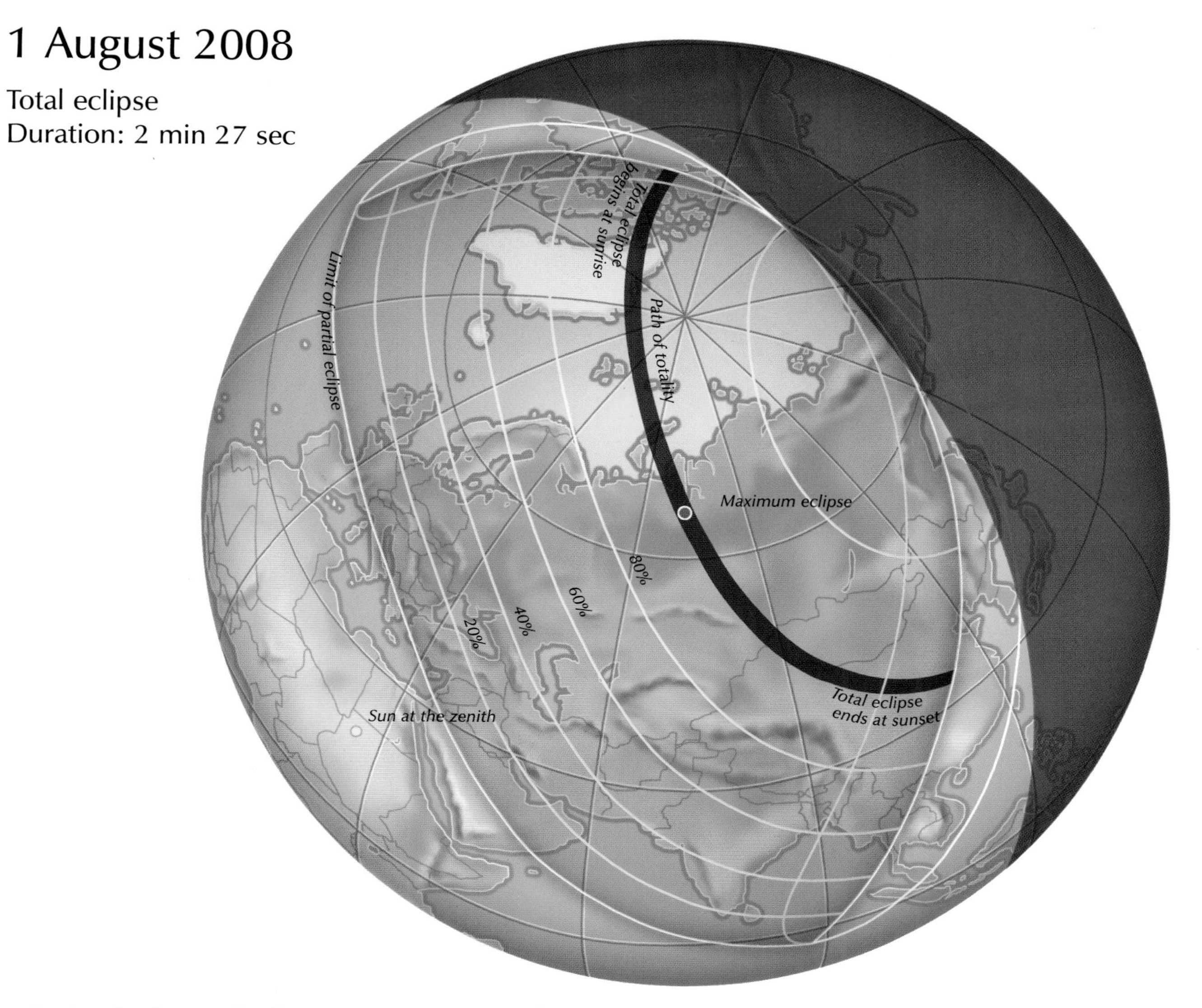

During the fine total eclipse on 1 August 2008, the Sun is in the constellation of Cancer. To the left of the hidden Sun, will lie the planets Venus, at 14°, and particularly Mercury, which is exceptionally close at just 3°30′. Only the end, at sunset, of the annular eclipse on 26 January 2009 is visible on solid land. The eclipse ends in Indonesia.

26 January 2009

Annualar eclipse
Duration: 7 min 53 sec

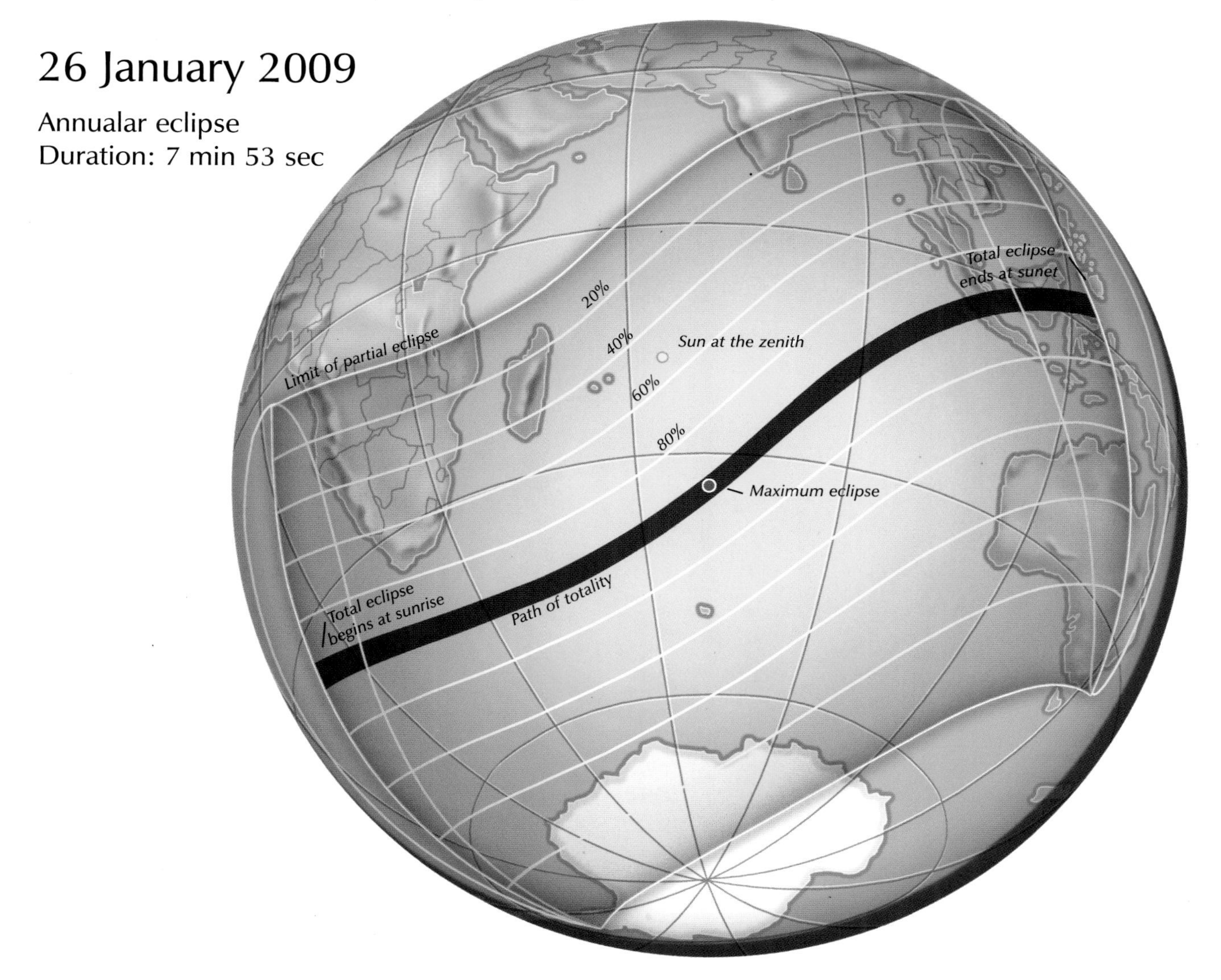

22 July 2009

Total eclipse
Duration: 6 min 38 sec

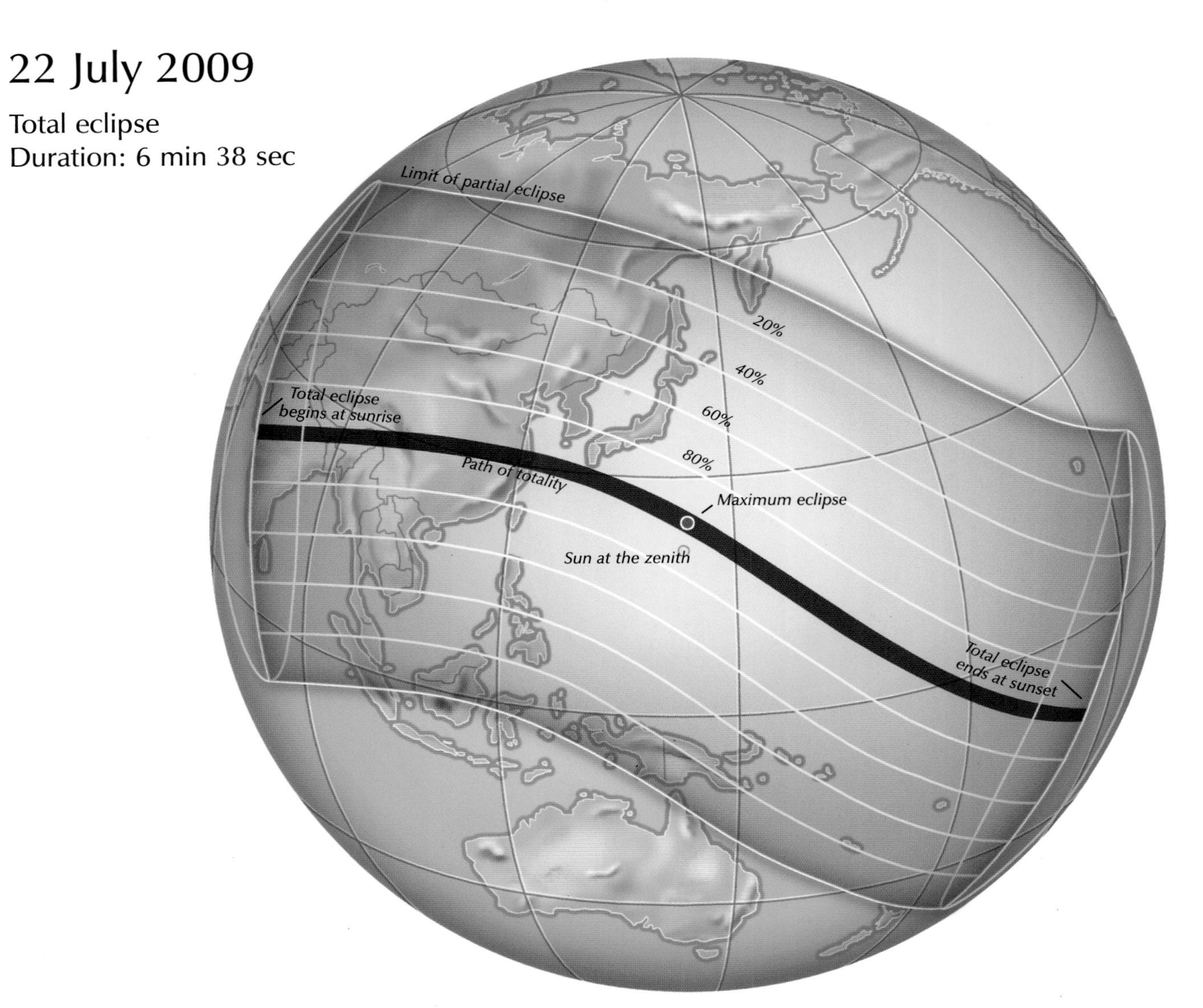

The eclipse of 22 July 2009 is one of the longest eclipses of the century. It begins on the western side of India, brushes past Nepal and Tibet, before crossing China from one side to the other. In the sky, Mercury will shine just 9° away from the Sun. The annular eclipse of 15 January 2010 is also exceptionally long, and appears in the sky over Africa, India and Asia.

15 January 2010

Annular eclipse
Duration: 11 min 7 sec

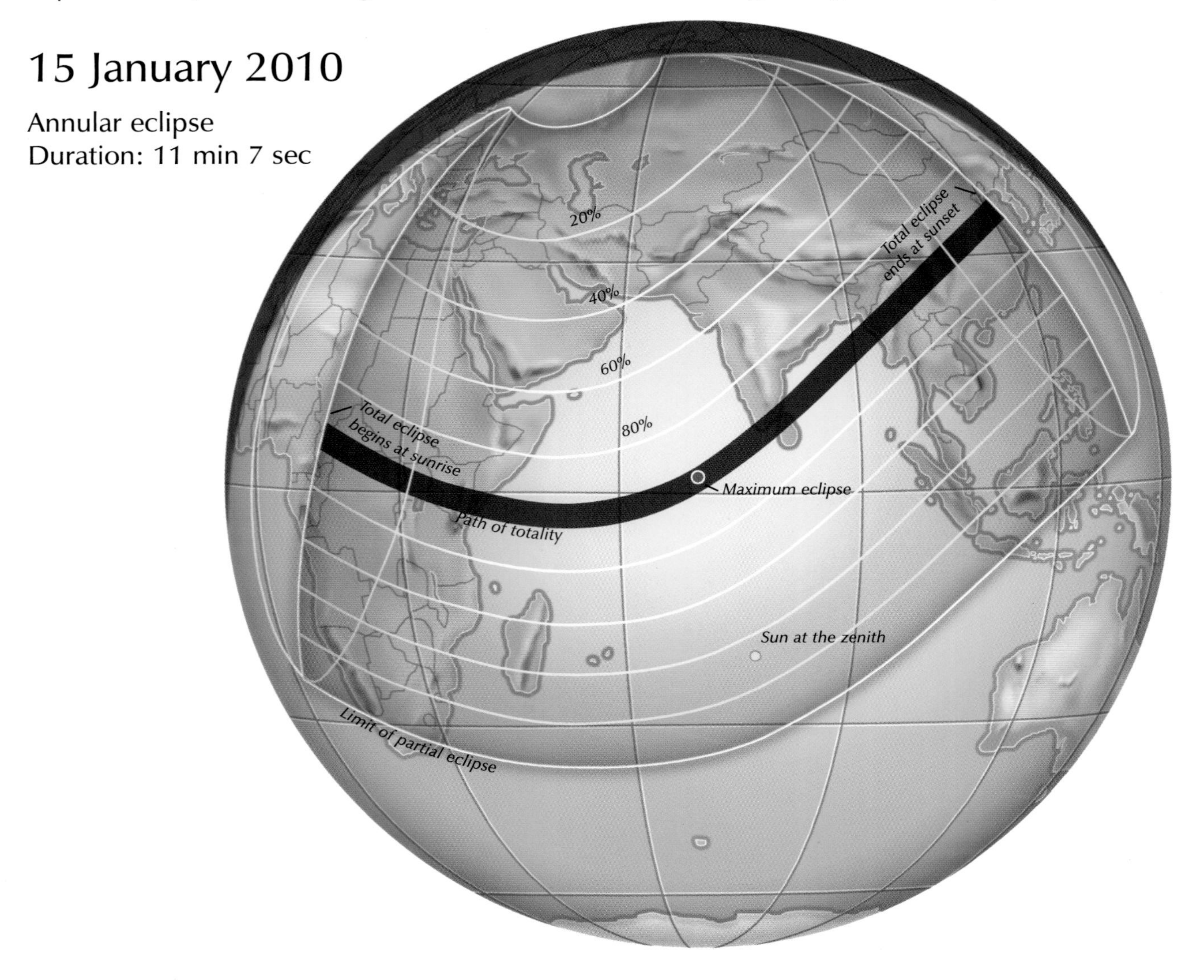

11 July 2010

Total eclipse
Duration: 5 min 20 sec

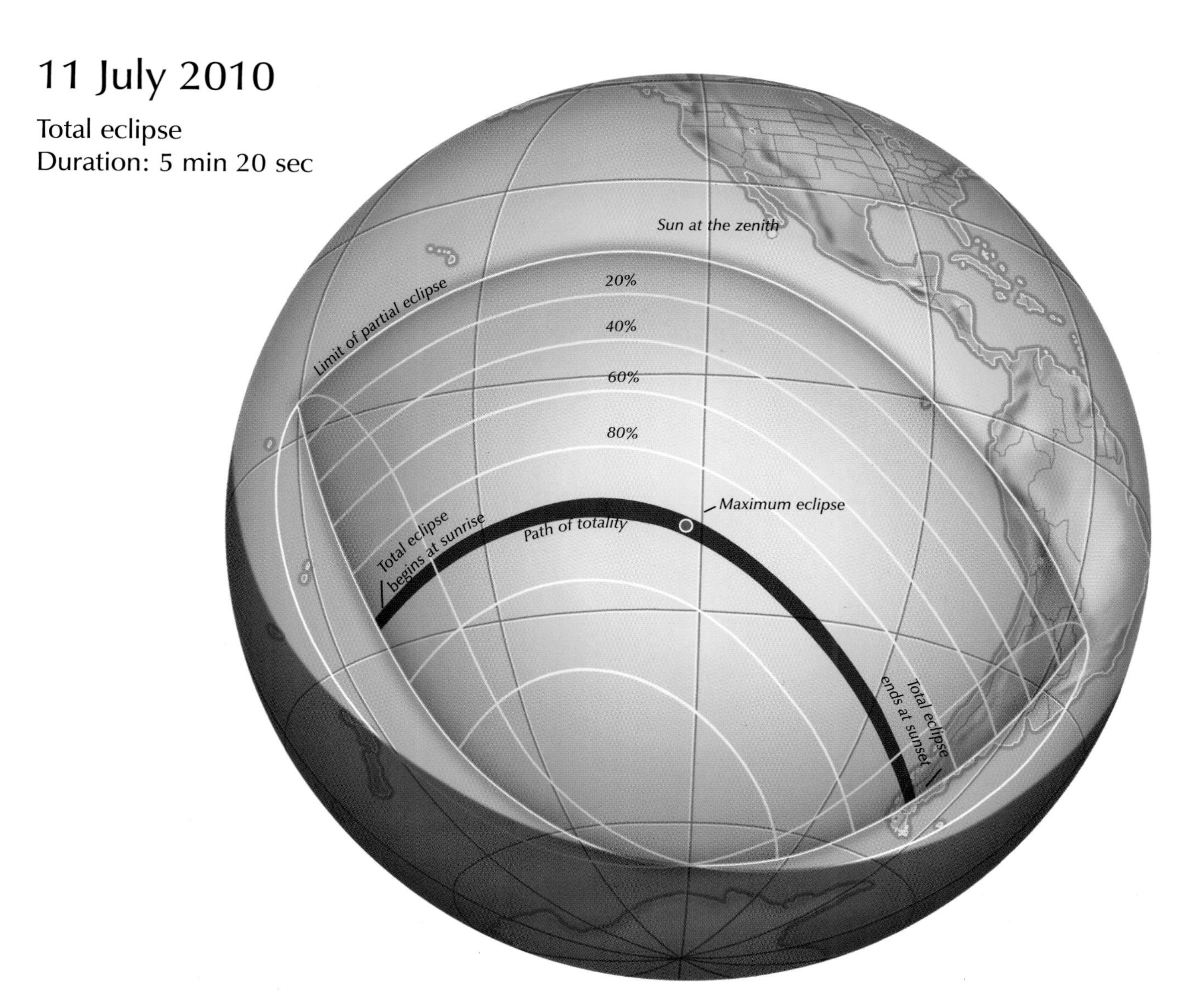

On 11 July 2010, the eclipse, which crosses the Pacific Ocean, will have few witnesses, except, perhaps, the gamblers who happen to risk being on Easter Island or the coast of Chile. Unfortunately, the middle of the southern winter means that observing this eclipse will be problematic. On 20 May 2012, the annular eclipse crosses the beautiful American deserts and the Grand Canyon at sunset.

20 May 2012

Annular eclipse
Duration: 5 min 46 sec

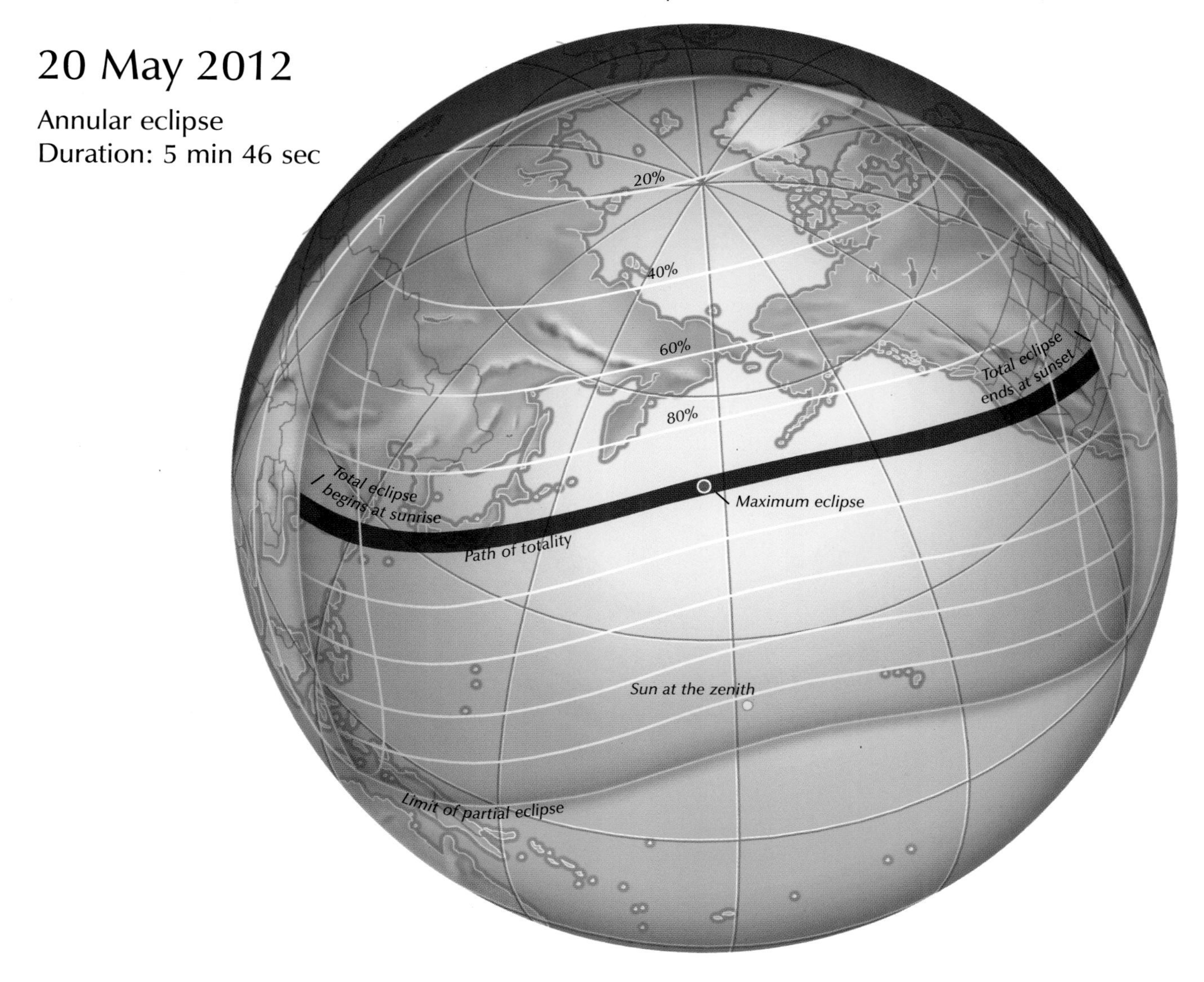

13 November 2012

Total eclipse
Duration: 4 min 2 sec

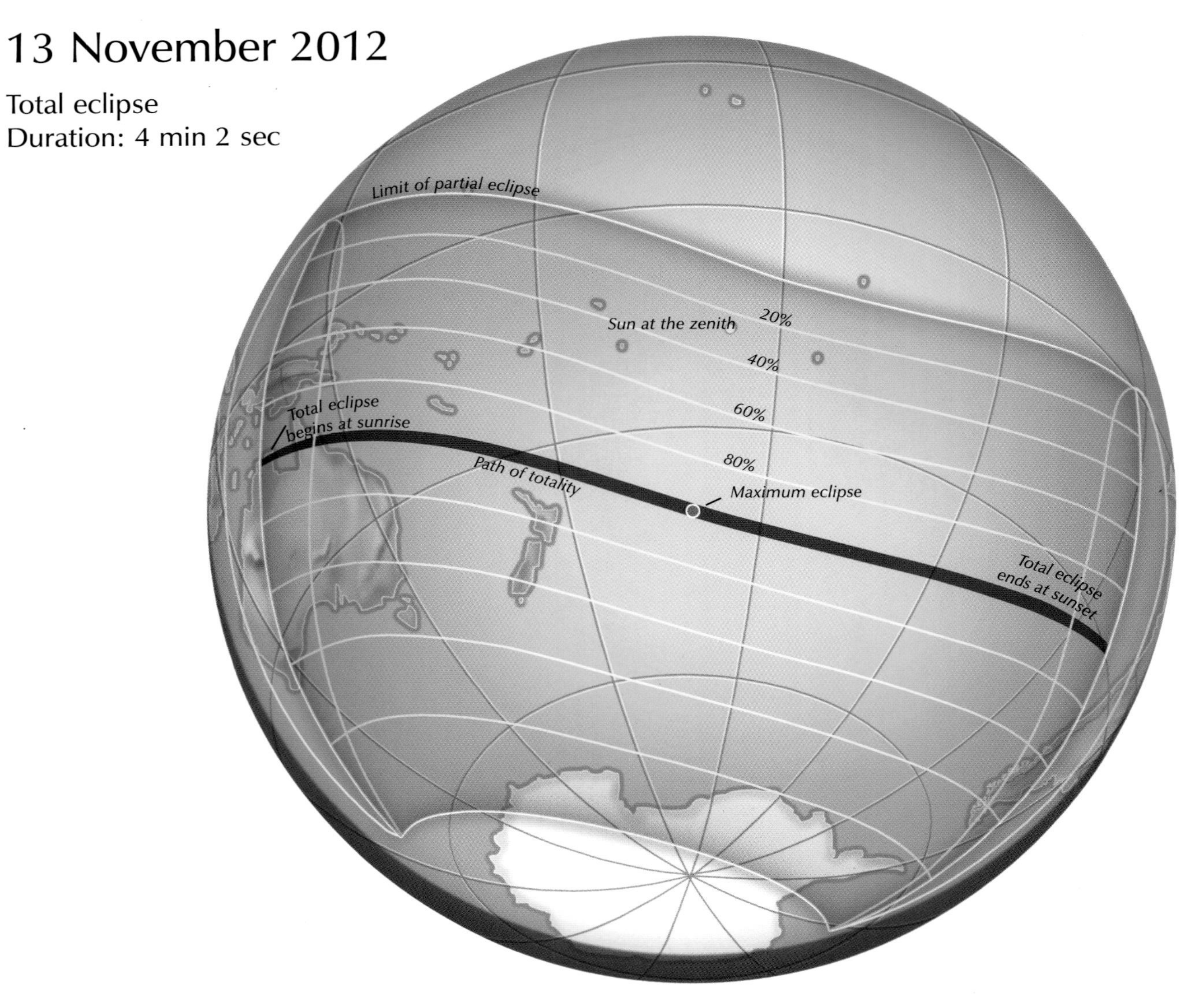

Saturn, Mercury and Antares in Scorpius surround the Sun during the eclipse on 13 November 2012. At Cairns, in Northern Australia, the eclipse lasts only 2 minutes. The Sun, which will have just risen, will lie 13° above the horizon. On 10 May 2013, Australia again sees an annular eclipse, which begins at sunrise in the Gibson Desert.

10 May 2013

Annular eclipse
Duration: 6 min 3 sec

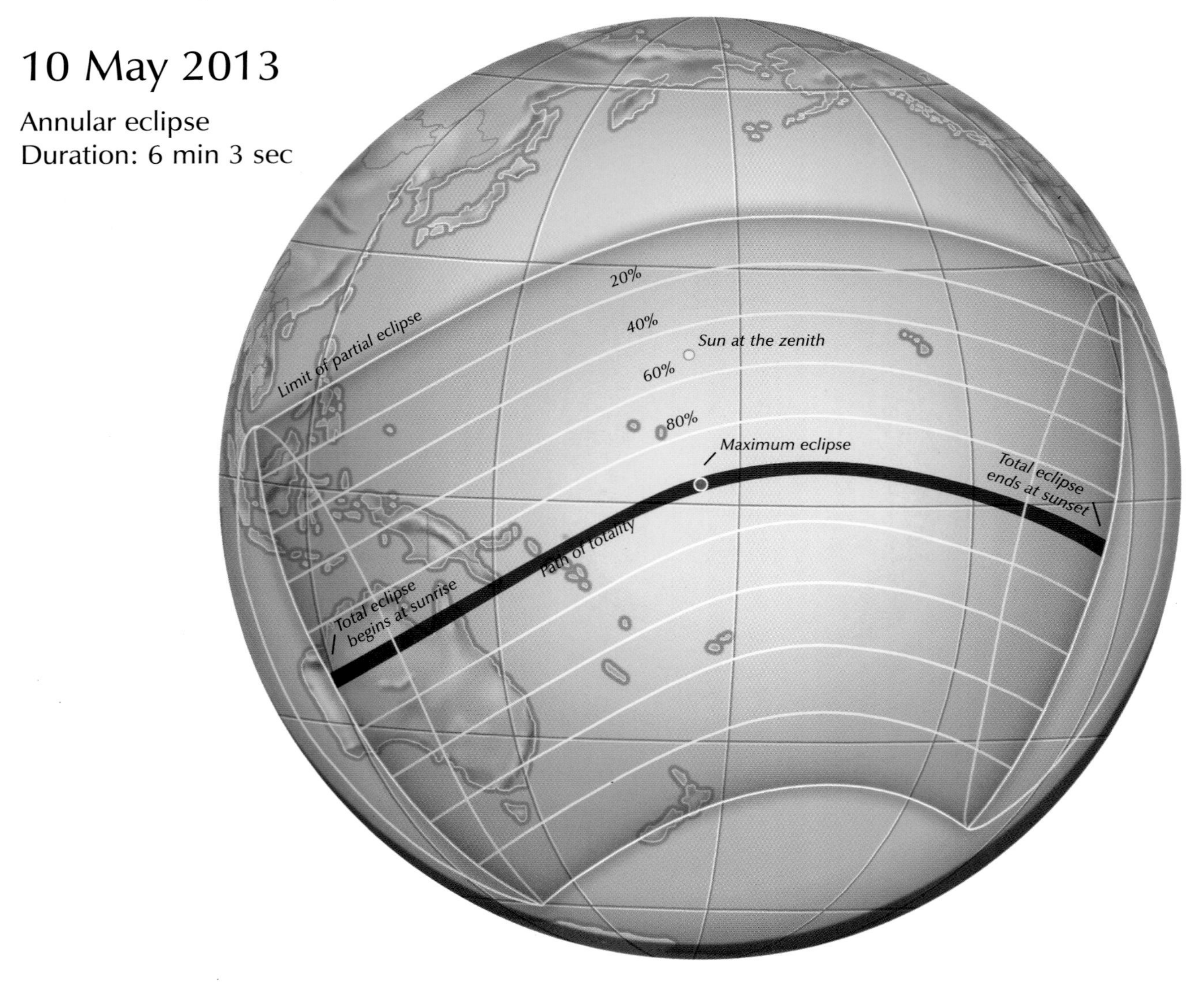

3 November 2013

Annular/total eclipse
Duration: 1 min 39 sec

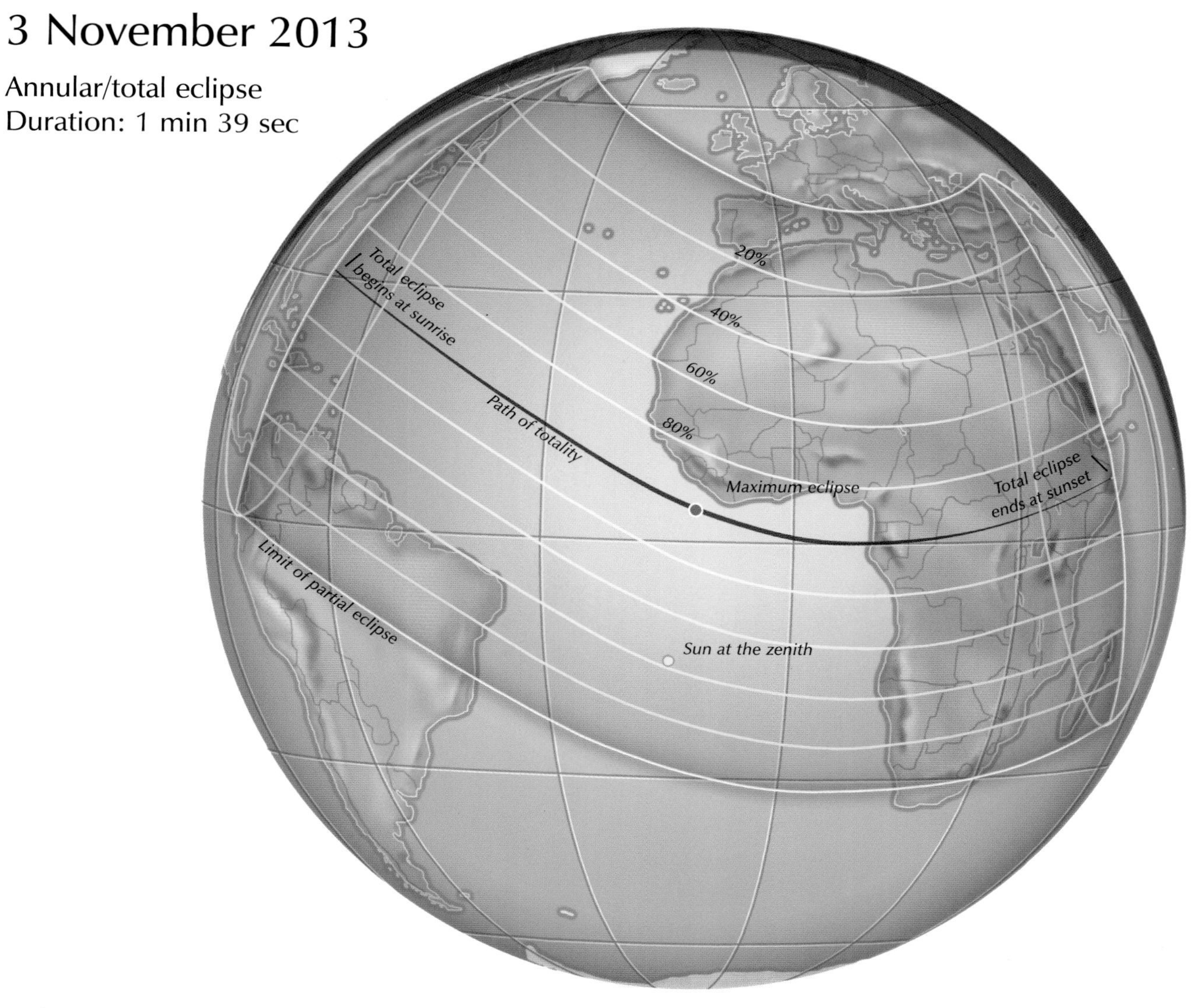

The eclipse of 3 November 2013 begins and ends as annular, but is total along nearly all of its path. On 20 March 2015, a fine total eclipse occurs at sunset at the North Pole. By an extraordinary chance, the eclipse takes place on the first day of sunlight at the North Pole, after six months of winter darkness! An eclipse to be admired from the Faeroe Islands or Spitzbergen.

20 March 2015

Total eclipse
Duration: 2 min 46 sec

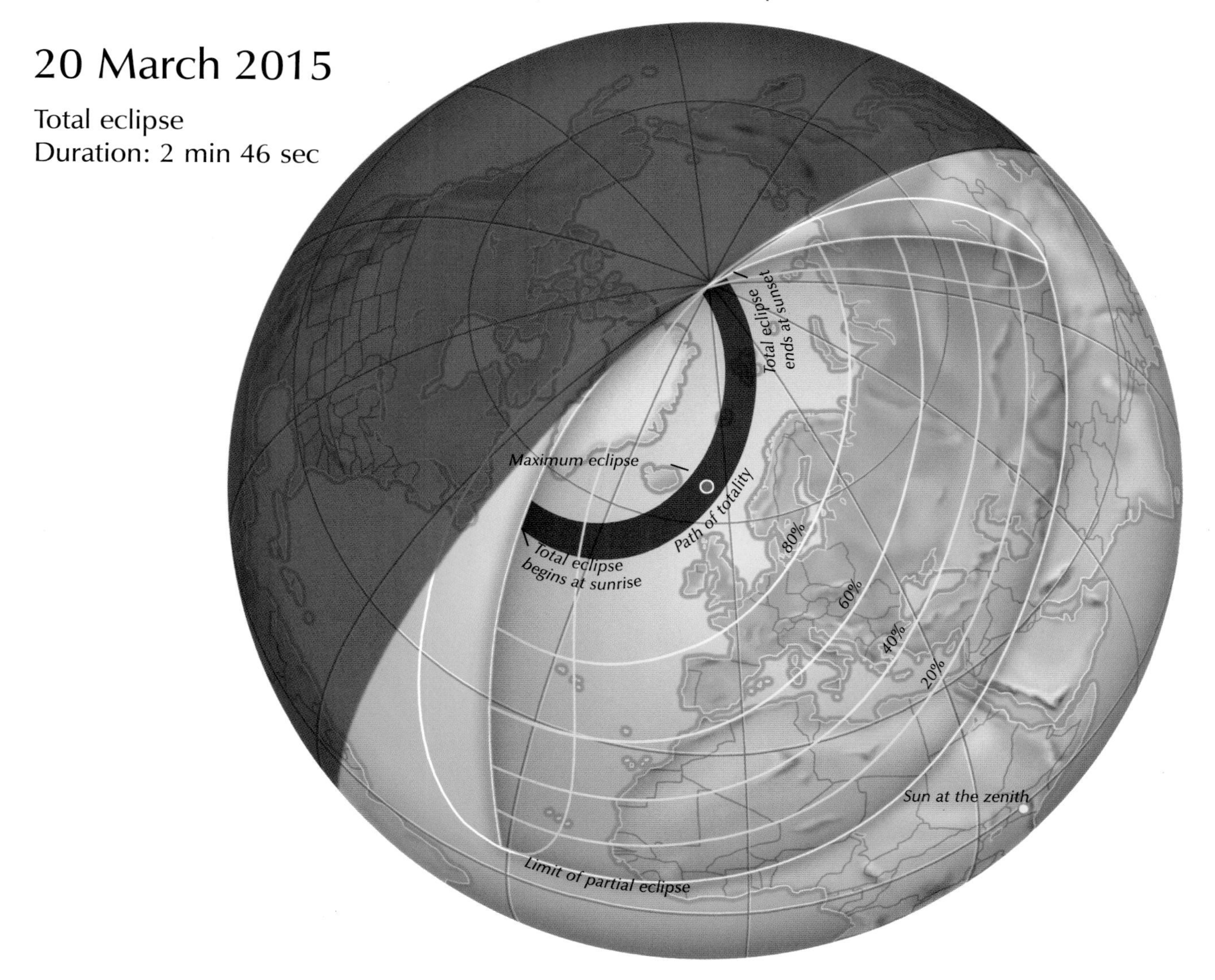

9 March 2016

Total eclipse
Duration: 4 min 9 sec

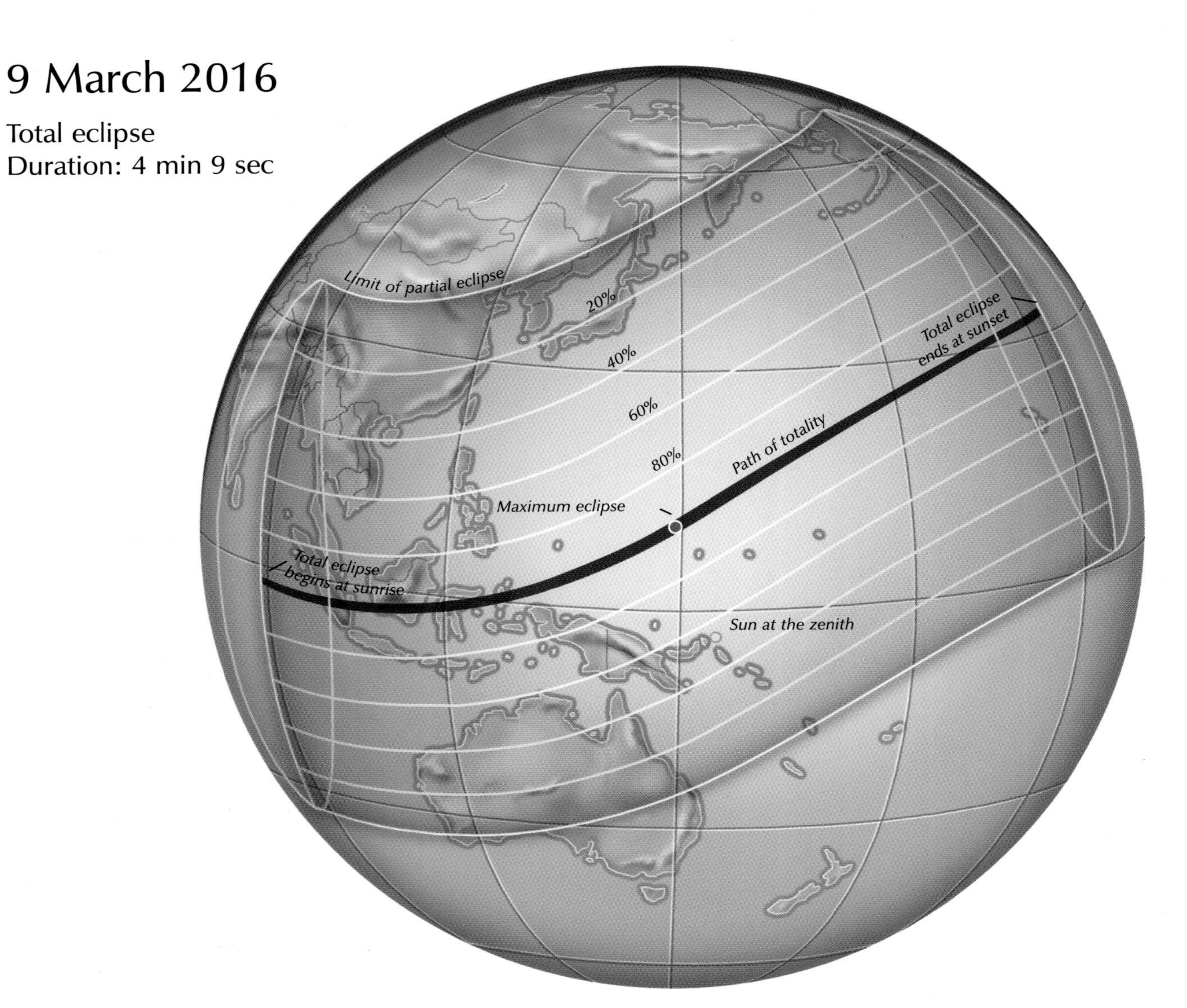

Venus and Mercury will appear in the darkened sky during the eclipse of 9 March 2016, visible in the early morning from Sumatra, Borneo, and Celebes. The eclipse then races across the Pacific. Beginning in the middle of the Atlantic, the annular eclipse of 1 September 2016 crosses Gabon, Zaire, Tanzania, Mozambique, and finally the large island of Madagascar.

1 September 2016

Annular eclipse
Duration: 3 min 5 sec

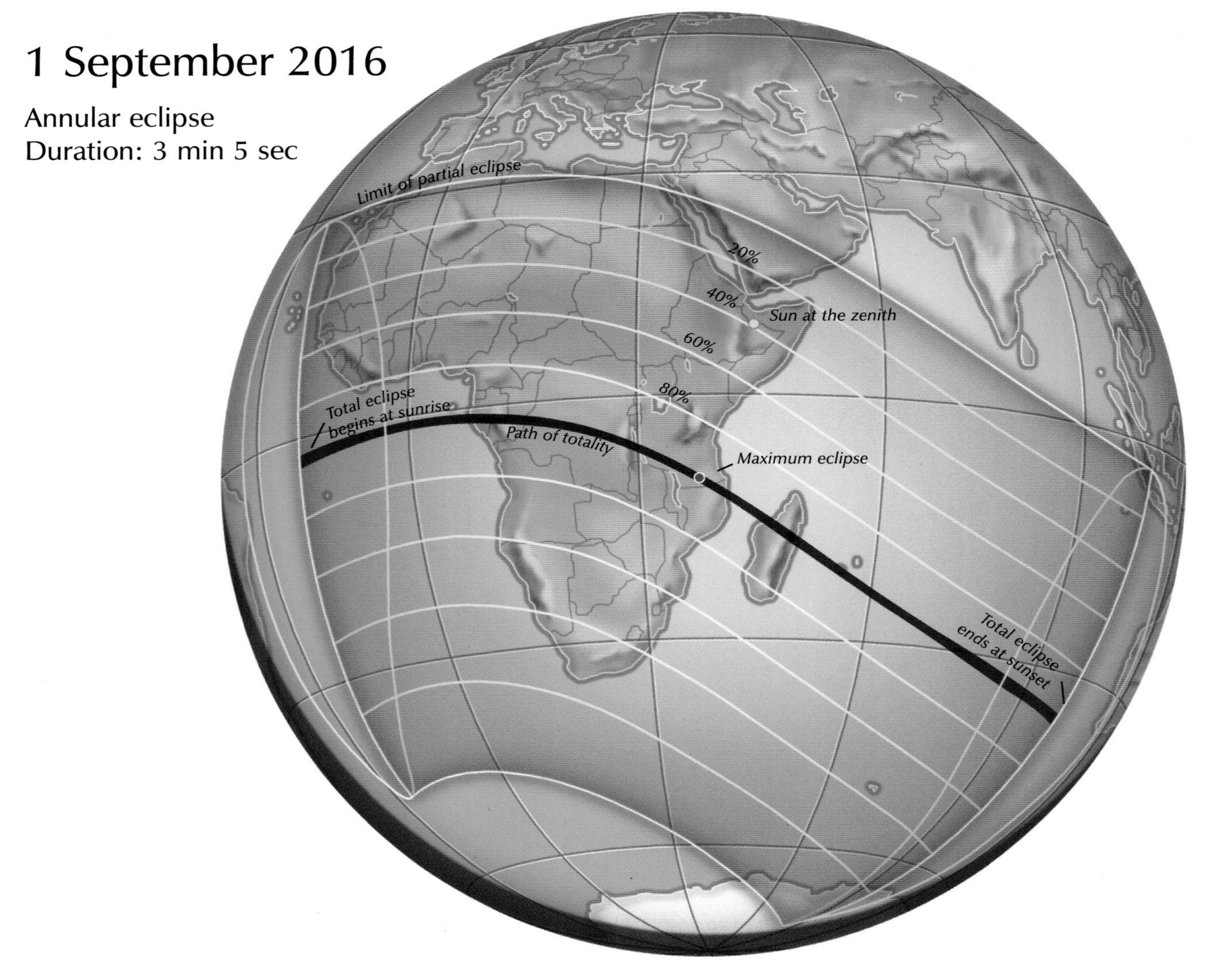

26 February 2017

Annular eclipse
Duration: 44 sec

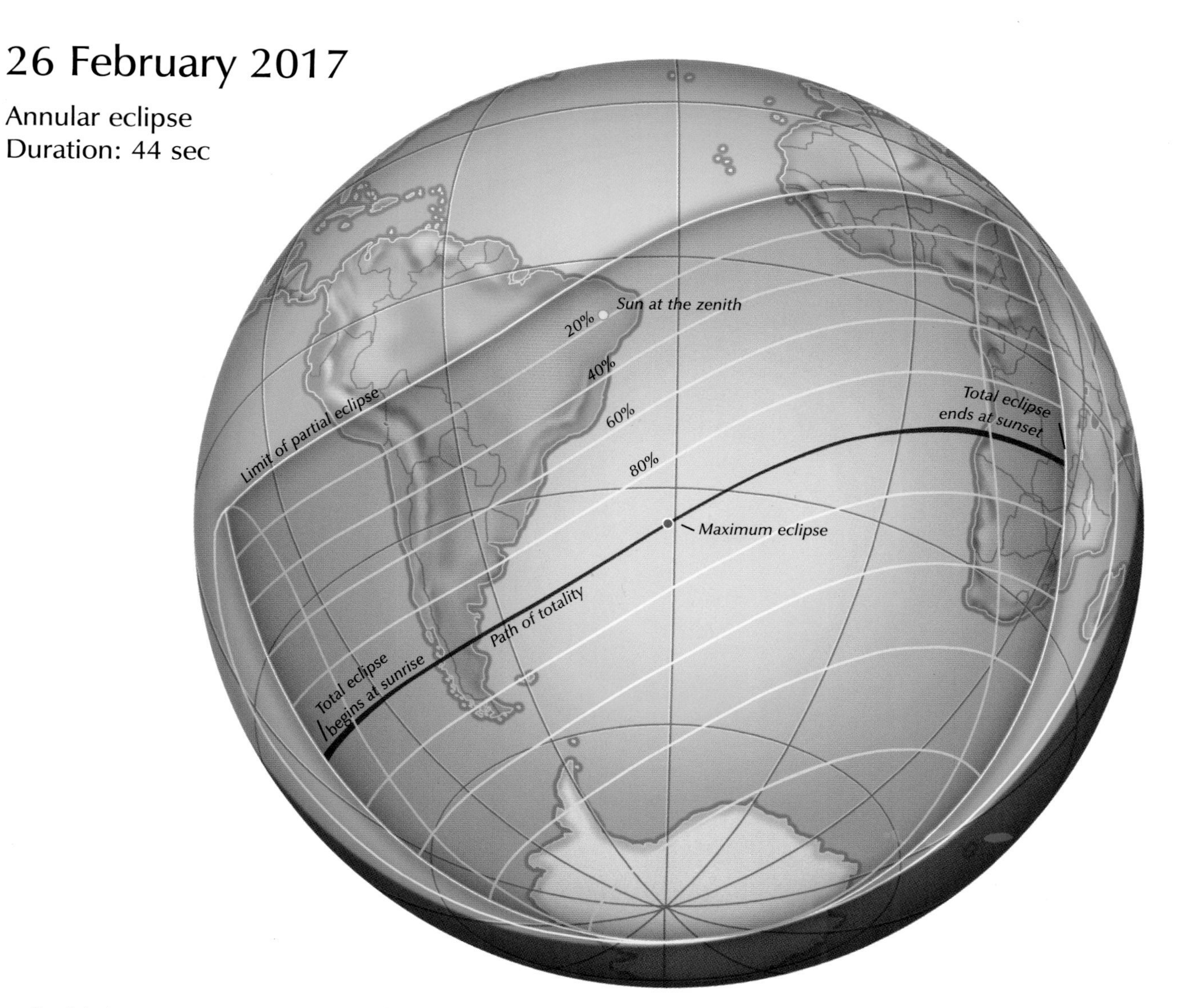

On 21 August 2017, an exceptional total eclipse crosses the United States. This eclipse is unique; it offers the chance to admire the beautiful star Regulus in Leo shining through the silvery wisps of the solar corona. Regulus will actually be just 1° away from the eclipsed Sun! Slightly farther away, the planets Mercury and Mars will frame the Sun and Moon.

21 August 2017

Total eclipse
Duration: 2 min 40 sec

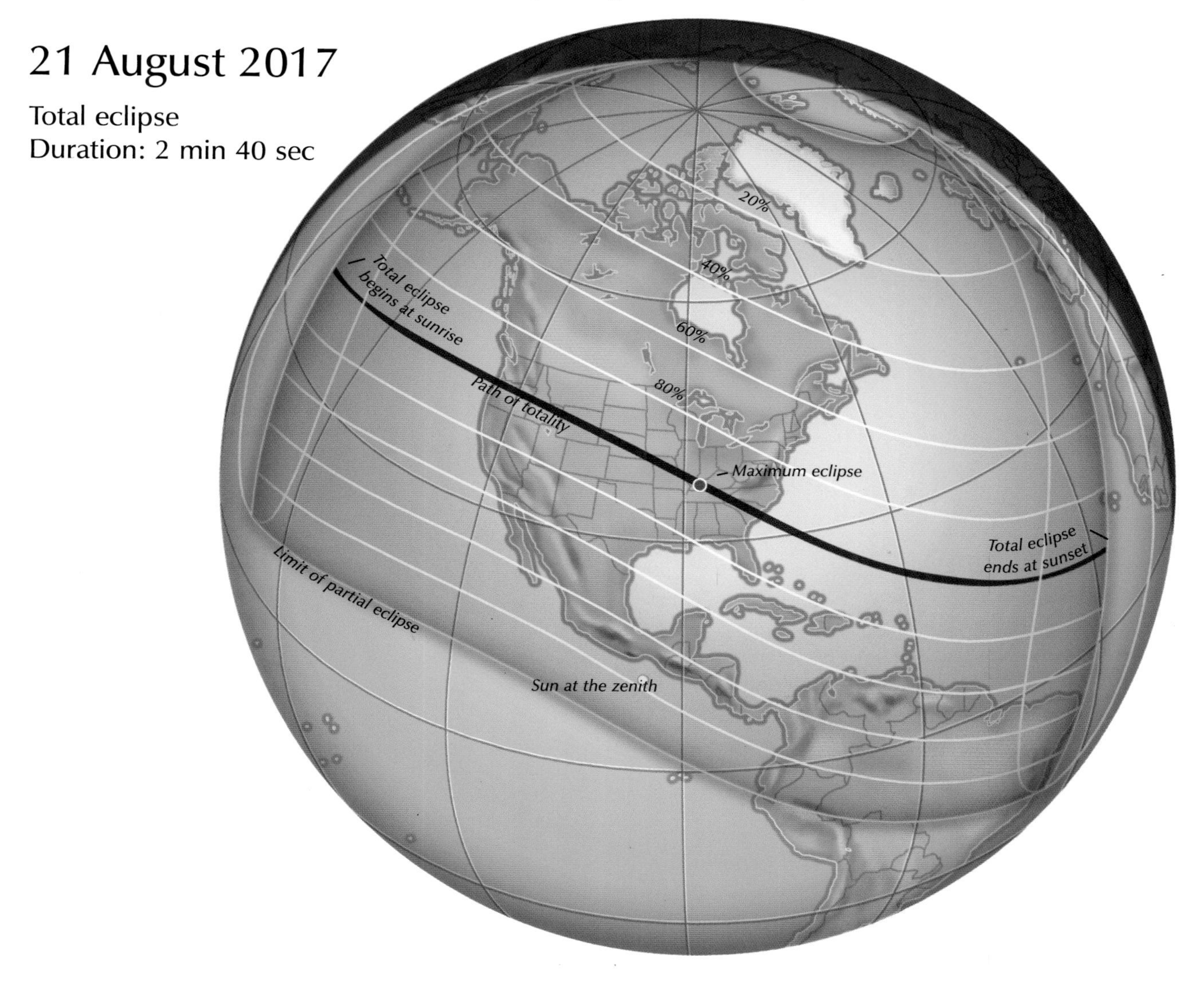

2 July 2019

Total eclipse
Duration: 4 min 32 sec

This is a magnificent total eclipse which crosses the Chilean coast, and then the Cordillera of the Andes, and finishes at sunset over Argentina. During this eclipse, on 2 July 2019, the eclipsed Sun will be in Gemini, surrounded by the planets Mercury, Mars, and Venus, and the stars Castor and Pollux, all visible to the naked eye during the total phase.

26 December 2019

Annular eclipse
Duration: 3 min 39 sec

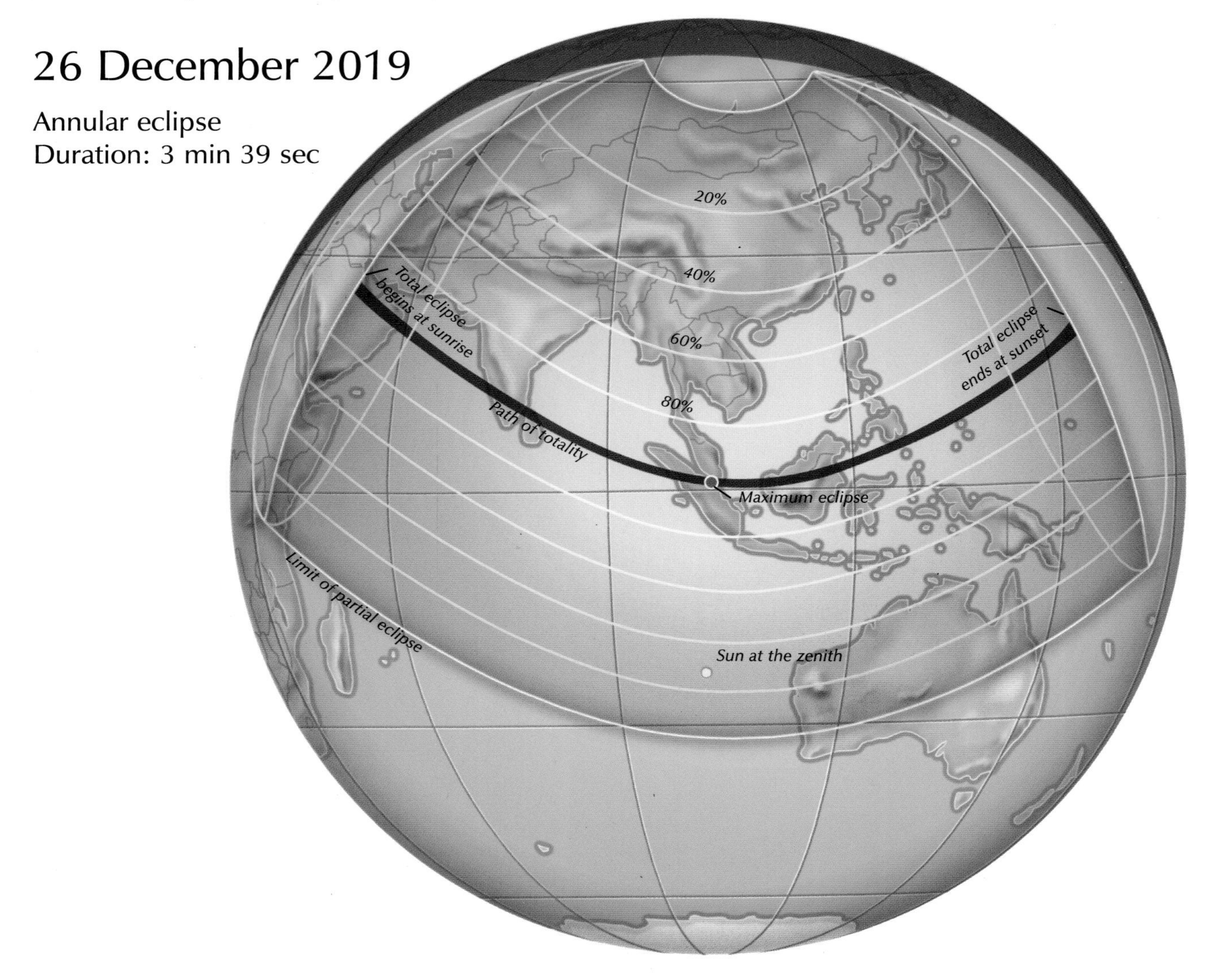

21 June 2020

Annular eclipse
Duration: 38 sec

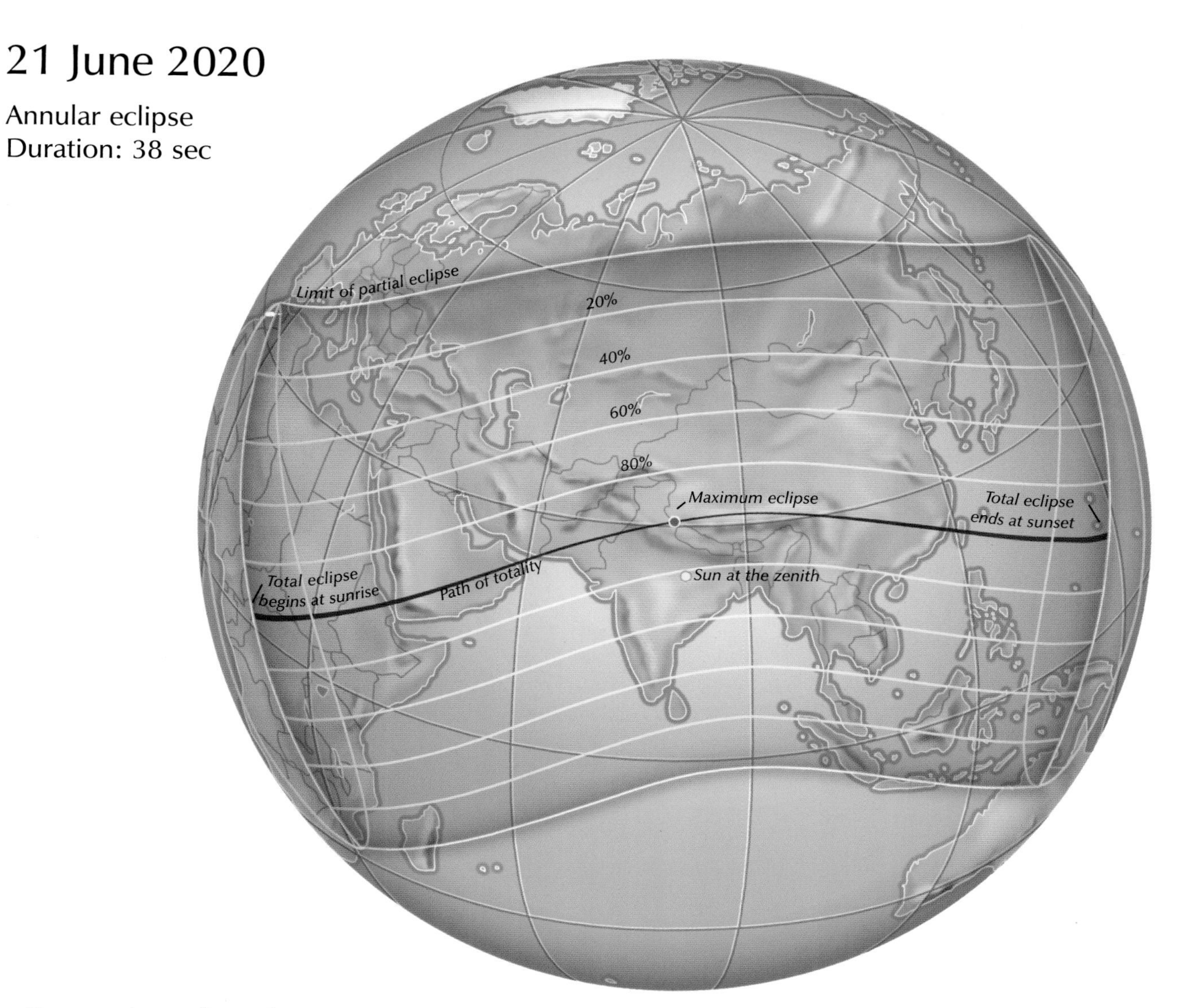

The very short eclipse of 21 June 2020, is visible practically at the zenith from the giddy peaks of the Himalaya, on the borders of India and Nepal. Six months later, on 14 December 2020, a fine total eclipse, plunges southern Chile and Argentina into night, at the end of the southern spring. Mercury, Antares in Scorpius and Venus will frame the eclipsed Sun.

14 December 2020

Total eclipse
Duration: 2 min 9 sec

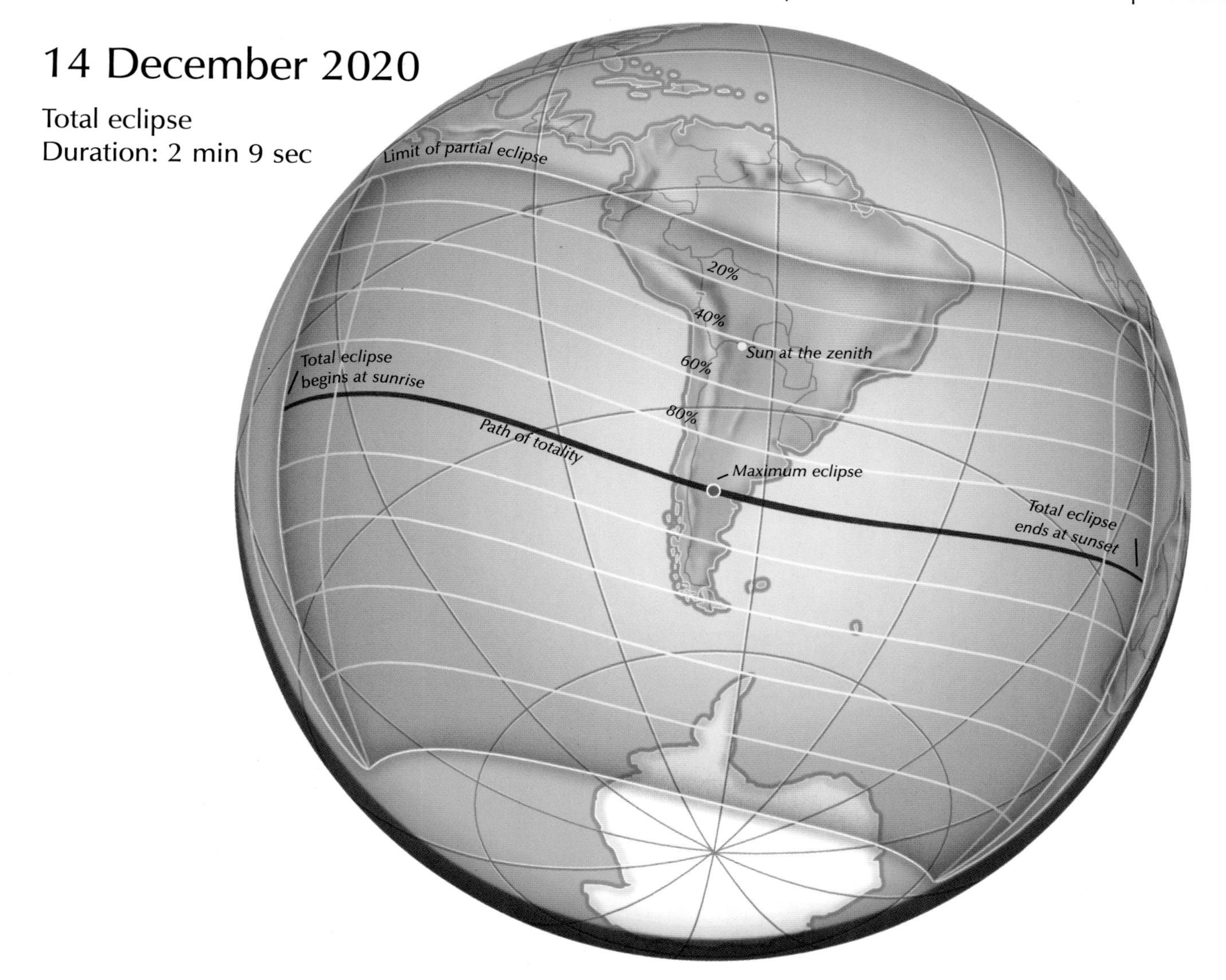

■ THE FULL MOON RISES AT THE SAME TIME AS THE SUN SETS. SOMETIMES, WHEN THERE IS A TOTAL ECLIPSE OF THE MOON, ITS ROSE-COLOURED TINT MINGLES WITH THE SHADES OF TWILIGHT.

It is the night of the Full Moon. The lunar disk, with the darker patches that dot its surface, inevitably suggests the surprised, whitened face of a clown. The Moon is brilliant, almost blinding. The shadows have receded in the face of so much light, leaving a fine night, with a dark blue sky, scattered with a few stars, and with enough light to read an astronomy book... And then, towards midnight, an enormous dark shadow slowly starts to eat away at disk of the Moon. As the minutes pass, it engulfs the highlands and the lava seas – a total eclipse has begun. Many readers will doubtless have already seen this event once or even twice. This is because, although there are more or less as many lunar eclipses as solar ones, the former, being visible from precisely half of the planet, are far more common than the latter. So, for the beginning of the 21st century, total eclipses of the Moon occur on 21 January 2000, 9 January 2001, 16 May 2003, 9 November 2003, 4 May 2004, 28 October 2004, and 3 March 2007. So they show none of the periodicity of total solar eclipses. Although less impressive than the latter, total eclipses of the Moon are magnificent astronomical spectacles. As with eclipses of the Sun, the best instruments for observing eclipses of the Moon are the naked eye, and binoculars. Amateur astronomers with refractors or reflectors can, of course, turn their instruments onto the Moon during an eclipse, but they should use them with the lowest possible magnification (let us say between 20 x and 80 x maximum), so that they can enjoy the sight of the coloured Moon in a field scattered with stars.

Naturally, observation of the Moon, unlike that of the Sun, does not involve any element of danger, and a filter is not essential, although to tell the truth, observing the Full Moon for a long time at the beginning of an eclipse, through a large instrument, can be rather unpleasant, because it is so bright. As with solar eclipses, the various initial phases of a total lunar eclipse are simply a 'taster'. For about an hour, as the Moon passes into the Earth's shadow, it slowly goes through all the phases.

Whether they are solar or lunar, eclipses are strictly predictable, to within a few seconds, centuries in advance. Yet no lunar eclipse is like any other, and the phenomenon always offers something new – whence its magic. Intuitively, the reader might imagine that once it has plunged deep into the Earth's shadow, the Moon simply becomes invisible. As we shall see, this is far from the truth. Let us, for a moment, imagine that we have been transported to the Moon to admire the eclipse from another point of view. As seen from the lunar surface, it is certainly an eclipse – of the Sun, because the Earth is between the Sun and the Moon. From the Moon, our planet appears as a

■ Full Moon on the evening of 27 September 1996. In a few minutes, our satellite will begin to enter the Earth's shadow. Its brightness will fall dramatically, and it will be decked in sunset colours.

■ In this photographic sequence of an eclipse, the Moon crosses the Earth's shadow from right to left. The centre image was exposed for much longer

black disk with an apparent diameter of 1°50′, i.e., about four times as large as the Sun's disk, which it covers completely, except for the outer corona. The large black disk of the Earth actually appears, during a solar eclipse, to be surrounded by a narrow, very bright, and orange-coloured halo. This halo consists of the terrestrial atmosphere, seen around the terminator, the whole circumference of the Earth where day is separated from night, and illuminated by both the last rays of the setting Sun and, on the other side of the world, by the first rays of dawn. The rays of light that have passed through the Earth's atmosphere have been partly absorbed, and also refracted as if through a prism, so that they finally reach the surface of the Moon, which they illuminate in a ghostly fashion.

This is why, when seen from the Earth, the Moon appears to be coloured during the middle of the eclipse. What makes total eclipses of the Moon so impressive and intriguing is that it is extremely difficult to predict what colour and how bright they are likely to be. The brightness and colour of the Moon as it passes through the Earth's shadow – which lasts on average an hour and a half – depend on the amount of cloud cover on Earth, on the transparency of the lower atmosphere, on the Earth-Moon distance, etc. As a result, almost anything is possible. During the darkest eclipses, the colourless Moon is almost lost against the black of the sky. At the other extreme, during the brightest eclipses, it shines brilliantly in a sky scattered with stars, and is cloaked in wonderful twilight colours, appearing a coppery hue, orange, or even blood red. You will get the full benefit of the colours through a pair of binoculars, rather than observing just with the naked eye.

In every case, the weak intrinsic light of the eclipsed Moon, and the subtleties of the colours that may be seen, lead us to suggest that lunar eclipses should be observed from sites that are protected as much as possible from pollution and from artificial lights. The countryside or, better still, up in the mountains, offers the best observing sites, with dust-free skies and no scattered light. Under such perfect conditions, at the moment when the event is at maximum, that is, when the Moon is in the very centre of the Earth's shadow, the night sky resumes its normal appearance. The faintest stars and the Milky Way are perfectly visible once more. The Moon then appears as an interloper in this serene sky. Its brightness may decrease in a spectacular fashion: between one thousand and one hundred thousand

THAN THE OTHERS, TO COMPENSATE FOR THE ENORMOUS DIFFERENCE IN BRIGHTNESS BETWEEN THE PARTIAL AND TOTAL PHASES.

times, depending on the eclipse! In a few, extremely rare, cases it is so dark that it cannot be seen by the naked eye, nor in binoculars, nor even in a telescope.

Portraits of eclipses

One might imagine that such variations in illumination do not make it any easier to take a photograph of a lunar eclipse. In fact, such an activity is generally reserved for astronomers, even though, as we shall see, ordinary photographers can try to include a beautiful eclipsed Moon is a nocturnal landscape. Enemy number one for lunar eclipse chasers is not the weather, as with total solar eclipses, which generally last just 1, 2, 3, or 4 minutes. During a total lunar eclipse, a photographer frequently has at least between half and one-and-a-half hours. No: the problem that the photographer faces is the lack of brightness in the subject. During a total lunar eclipse, one is effectively working under night-time illumination. In these circumstances, it is not sufficient for the photographer to simply make a longer exposure, as one might imagine. The apparent motion of the sky, caused by the rotation of the Earth, forbids one from exposing the film for too long, for fear of recording movement of the Moon and the stars on the final image. This movement is, of course, the greater the longer the focal length (and thus the resolution) of the lens being used (see Table p.174).

Any photographer who wants to tackle photographing an eclipse in a nocturnal scene – with a forest or snow-capped mountains as a foreground, for example – needs to have a lens with a short focal length (15–50 mm), and a wide aperture, giving a focal ratio of between f/1.2 and f/2.8. If the camera is mounted on a firm tripod, colour film of 100 to 200 ISO is capable of obtaining a spectacular image, provide the exposure time is carefully calculated (see Table p.174). Once again, the exposure times are by way of example. Atmospheric conditions and the eclipse's intrinsic brightness need to be taken into account when exposures are calculated. As with solar eclipses, it is advisable to bracket exposures on either side of these average values.

The eclipse through a telescope

The lack of light during lunar eclipses, which gives a problem with simple photography, becomes crucial for an astronomer who wants to obtain an image of our near neighbour using a telephoto lens, a refractor, or a reflector. By their very nature, these instruments are not very fast. Their maximum focal ratio generally does not exceed 5.6 or 8. Under these circumstances, it simply impossible to photograph the total phase without an equatorial drive, that is, without an astronomical mounting, equipped with a small electric motor, which compensates for the apparent rotation of the sky (one rotation per 24 hours). With such a mount, and provided that it is also very stable and accurate, it is possible to succeed in obtaining some spectacular images of total eclipses. The mounting, naturally, needs to be properly adjusted, with its rotation axis pointing either towards the true celestial pole, which is the projection of the Earth's rotation axis, or to the Pole Star, which is very close. Any amateur-astronomy book will give adequate advice about aligning equatorial mounts. There remains, however, one difficulty, that purists will not fail to investigate thoroughly. An equatorial mounting does indeed compensate for the apparent rotation of the sky, but the Moon has its own apparent proper motion, averaging 0.5″ per second of time. In six seconds it therefore moves by 5″, in one minute, 30″, etc. This motion, which one might consider negligible, may however cause a significant movement on images obtained with optics that have focal lengths of 1000–2000 mm or more. Experienced astrophotographers will therefore, if possible, use an equatorial mounting with a variable-speed drive, and will also put the time before the total phase to good use by ensuring that their telescope's speed of rotation matches that of the Moon. If this is the case, then exposures of one or two minutes are not unreasonable. They will allow one to record the coloured image of the Moon even when the latter, is plunged into darkness, and seems to have forsaken the night sky. ■

LUNAR ECLIPSES FOR PHOTOGRAPHERS

Aperture	1.8	2.8	4	5.6	8
Bright totality	1 s	2 s	4 s	8 s	15 s
Dark totality	8 s	15 s	30 s	60 s	120 s

To photograph a landscape under the dim lighting of a total lunar eclipse, all that is required is to use a standard camera back, and a lens with a short focal length and wide aperture. The camera should be set to manual mode. This table gives an indication of the exposure times to use, with 100 ISO film and various aperture values, to photograph the landscape during the total phase.

LUNAR ECLIPSES FOR ASTRONOMERS

Aperture	2.8	4	5.6	8	11
Partial phases	1/1000 s	1/500 s	1/250 s	1/125 s	1/60 s
Bright totality	1 s	2 s	4 s	8 s	15 s
Dark totality	8 s	15 s	30 s	60 s	120 s

This table gives typical exposure times, for 100-ISO film and different aperture ratios (f/D), for lenses or astronomical instruments, to be used when photographing the partial and total phases of a total lunar eclipse.

FIELD SIZES AND EXPOSURE TIMES FOR PHOTOGRAPHING THE MOON

Focal length	Field covered by 24 x 36 mm frame	Size of the Moon (average 30′)	Exposure time (maximum without trailing)
20 mm	60° x 90°	0.2 mm	45 s
28 mm	50° x 73°	0.3 mm	30 s
35 mm	38° x 55°	0.4 mm	20 s
50 mm	27° x 40°	0.5 mm	10 s
135 mm	10° x 15°	1.3 mm	4 s
180 mm	7° x 11°	1.5 mm	3 s
300 mm	4°30′ x 6°50′	2.8 mm	2 s
400 mm	3°30′ x 5°	3.7 mm	1 s

This table enables one to see the photographic field given by most normal and telephoto lenses, telephoto lenses, with the standard format of 24 ¥ 36\,mm. For guidance, we also give, in millimetres, the size of the image of the Moon on the film as a function of focal length. Finally, for photographers who do not have an equatorial mounting and who use a simple tripod, we have shown the maximum exposure time that may be used without motion being detectable in the image. To obtain the sharpest possible images, however, it is advisable to use the shortest exposure, whenever possible.

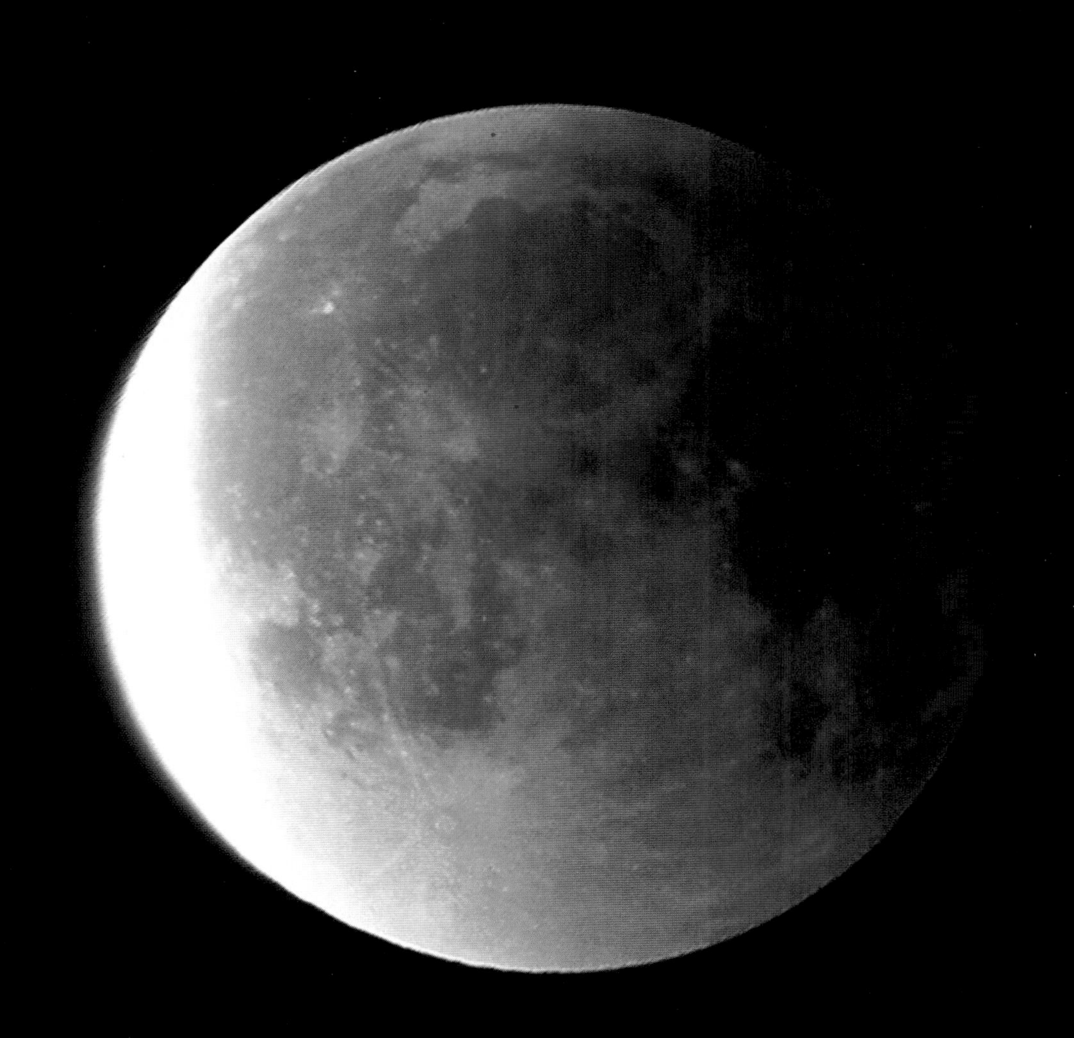

■ At the end of the night of 27 September 1996. The Moon is almost completely submerged in the Earth's shadow. Its brightness has fallen by a factor of ten thousand. Only the light from the Earth's atmosphere feebly illuminates the Queen of the Night.

Total eclipses of the Moon between 2000 and 2020

21 January 2000

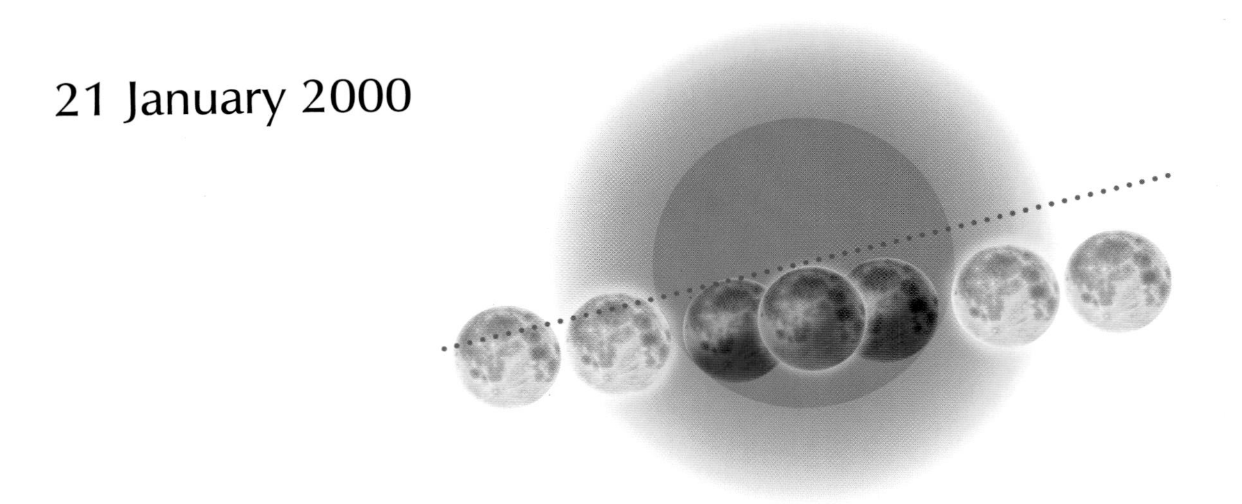

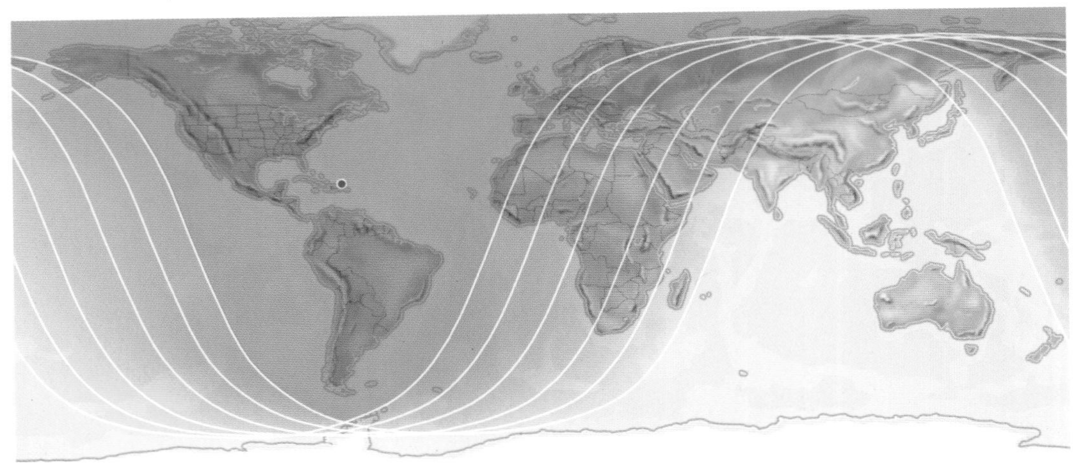

■ On these charts, the region of the Earth where the eclipse is total is shown by a darker tint – because it is plunged into darkness by the eclipse. From this region, all of the eclipse, including the partial phases, is observable.

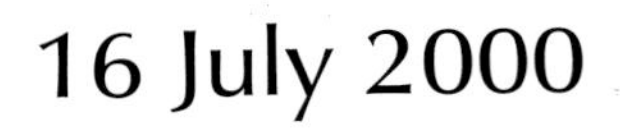

16 July 2000

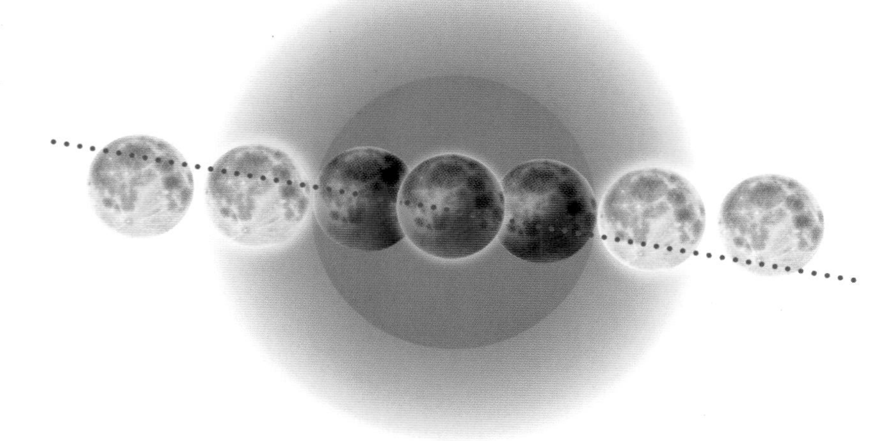

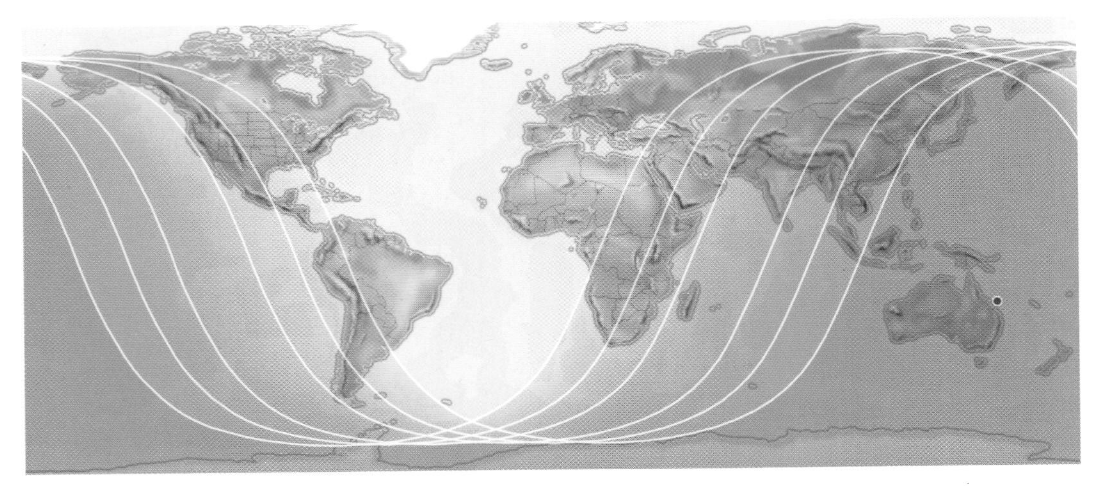

■ Visible only in China, Indonesia, Australia, and New Zealand, this is the very last total eclipse of the millennium. Almost central, the eclipse of 16 July 2000 will probably be very dark.

9 January 2001

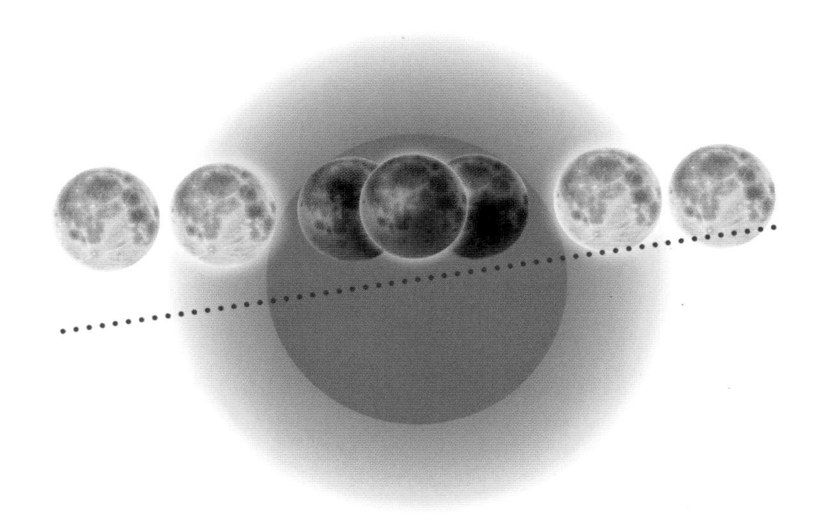

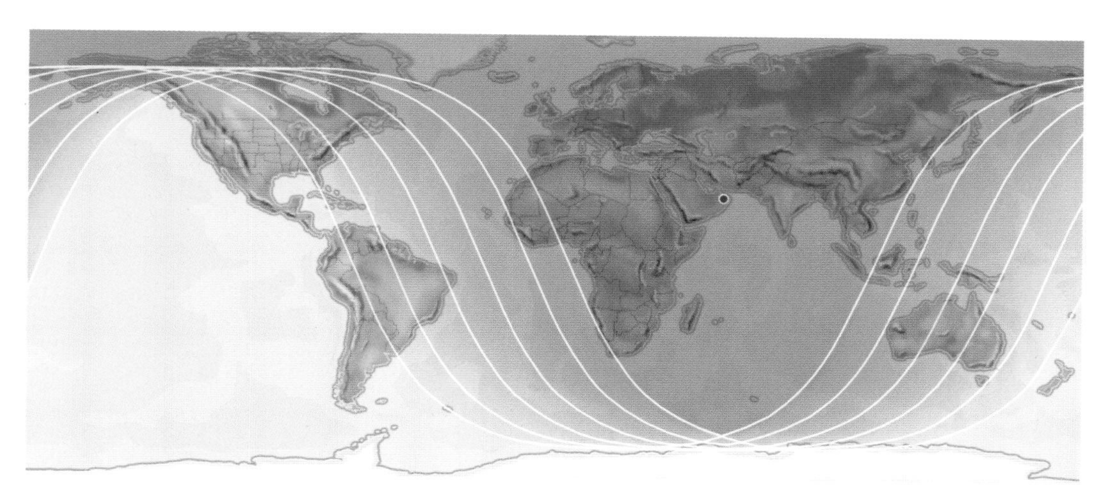

■ The third millennium opens with a magnificent total eclipse, visible from Europe, Africa, and Asia. During this winter eclipse, the Moon will be very high in the sky over Europe and North Africa.

9 May 2003

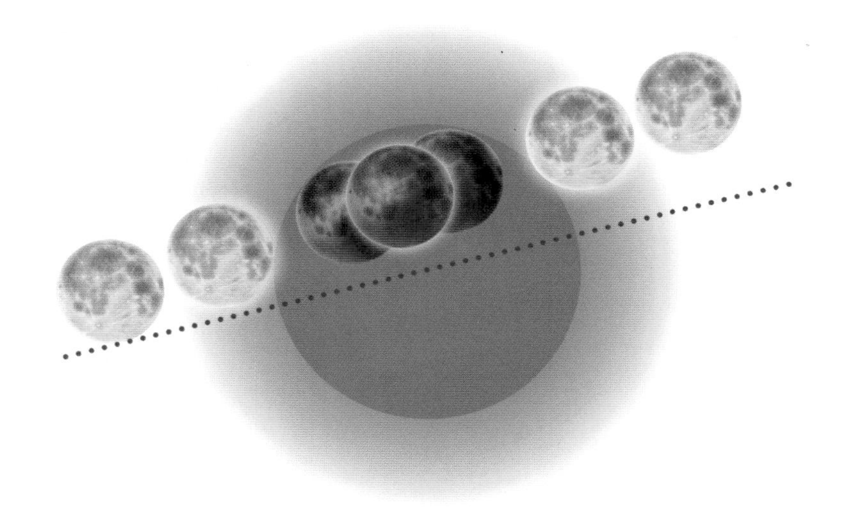

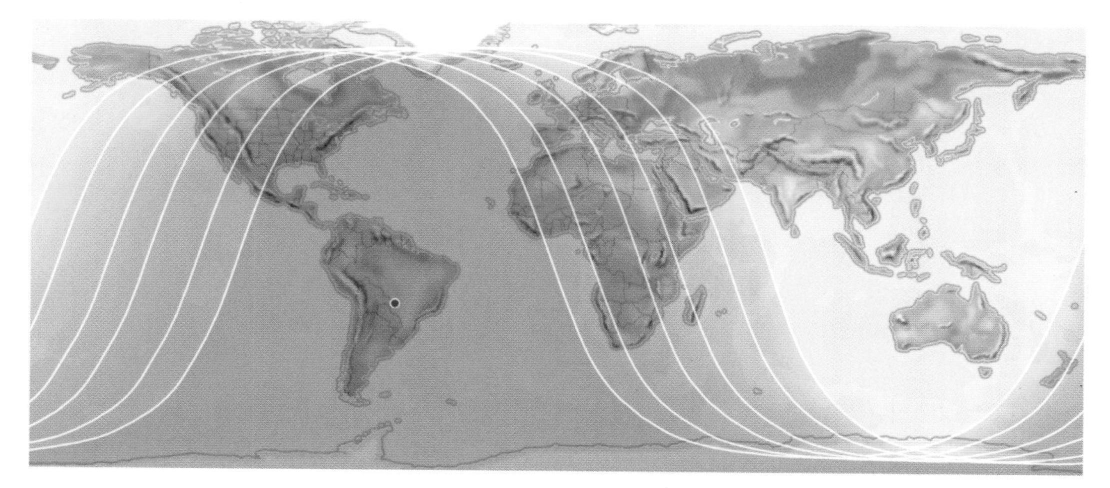

■ On either side of the region where all of the eclipse is visible, other countries see a part of the event. These regions are indicated by lighter and lighter graduated tints of grey.

9 November 2003

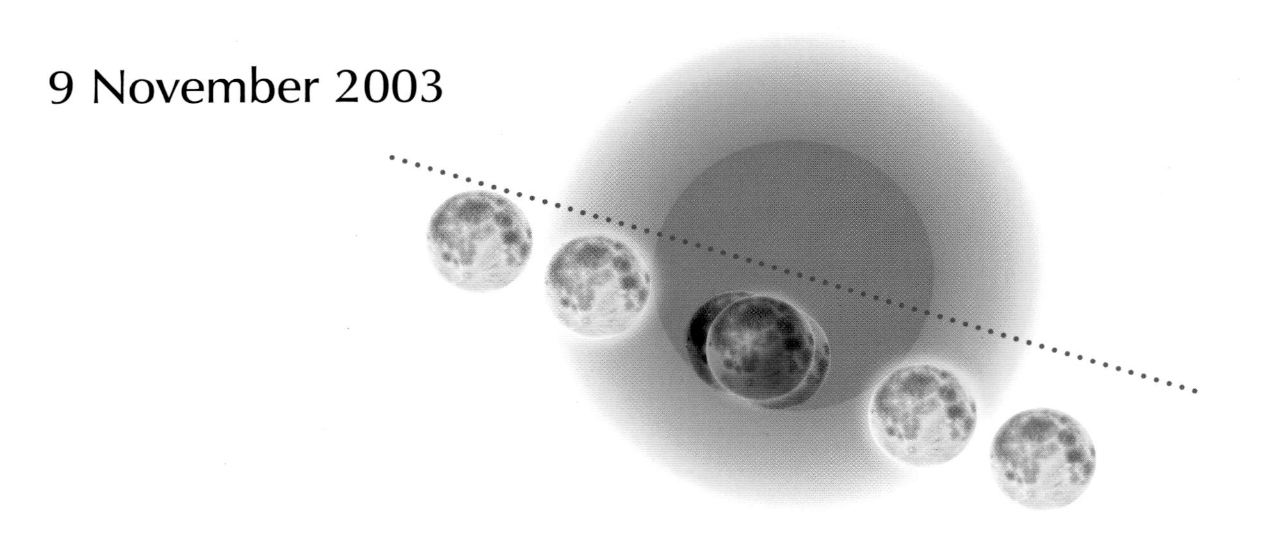

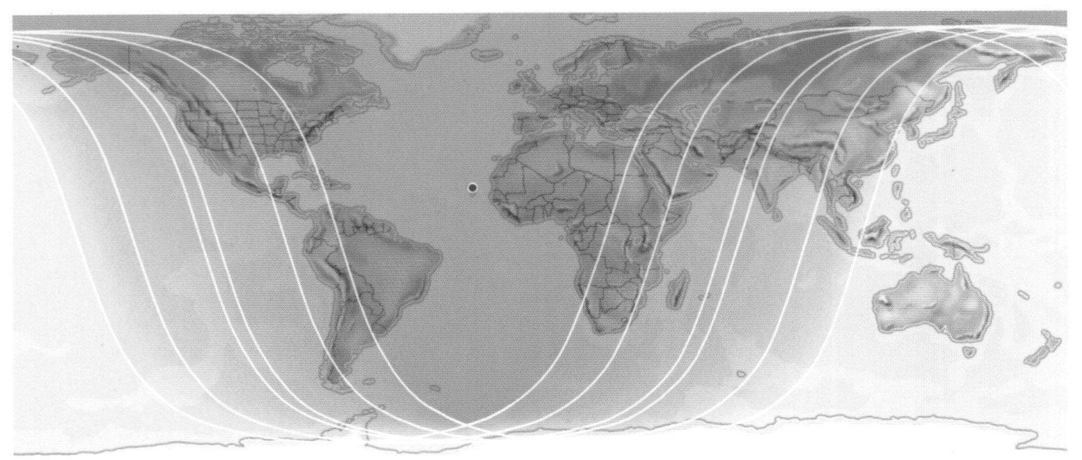

■ This eclipse should be bright: the Moon will brush the edge of the umbra throughout the event, which will be visible from the whole of Europe, West Africa, Brazil, and the eastern side of North America.

4 May 2004

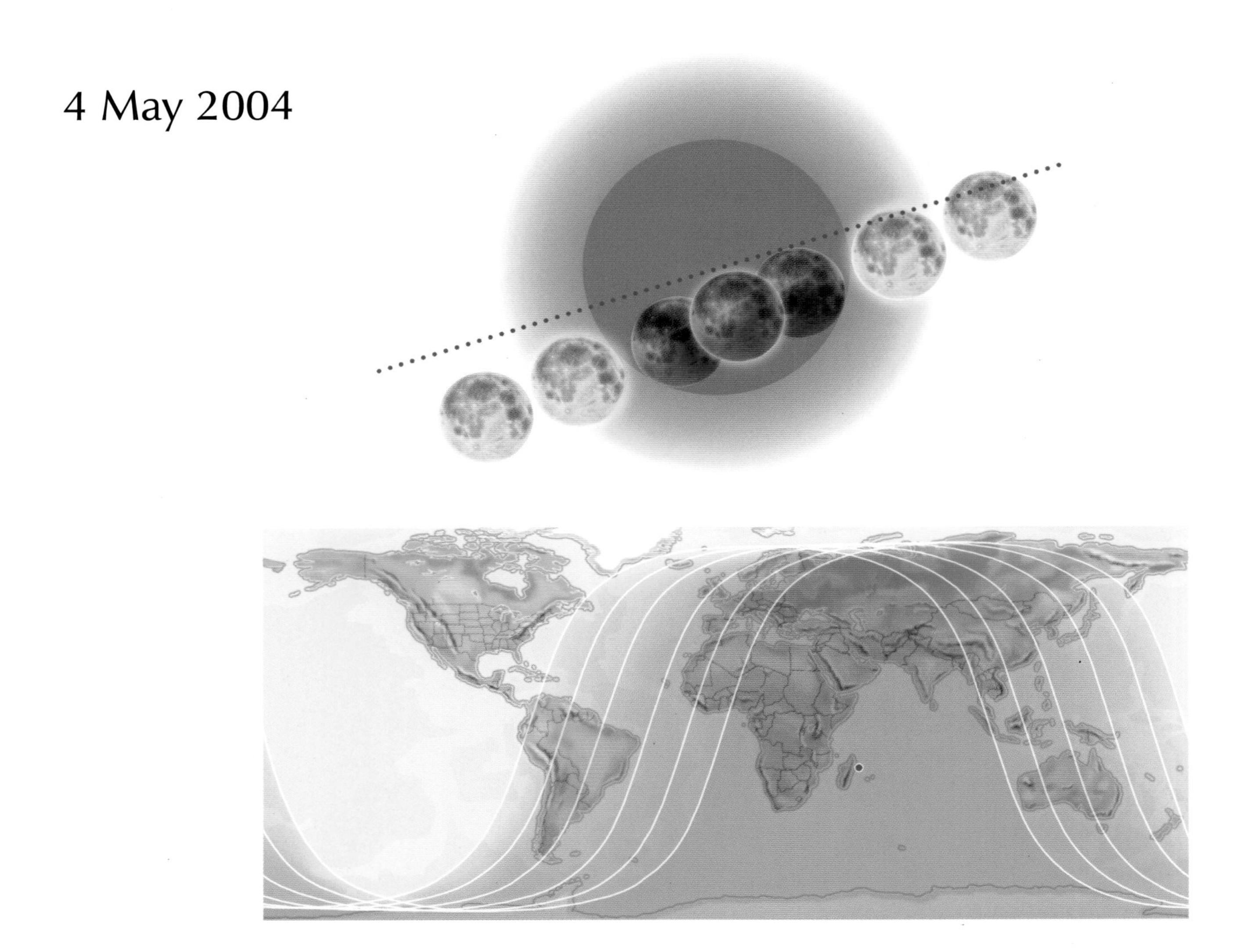

■ During this eclipse, the Moon will be close to the zenith over the islands of Madagascar, Mauritius, and La Réunion. On each chart, the point vertically beneath the Moon is indicated by a small dark circle.

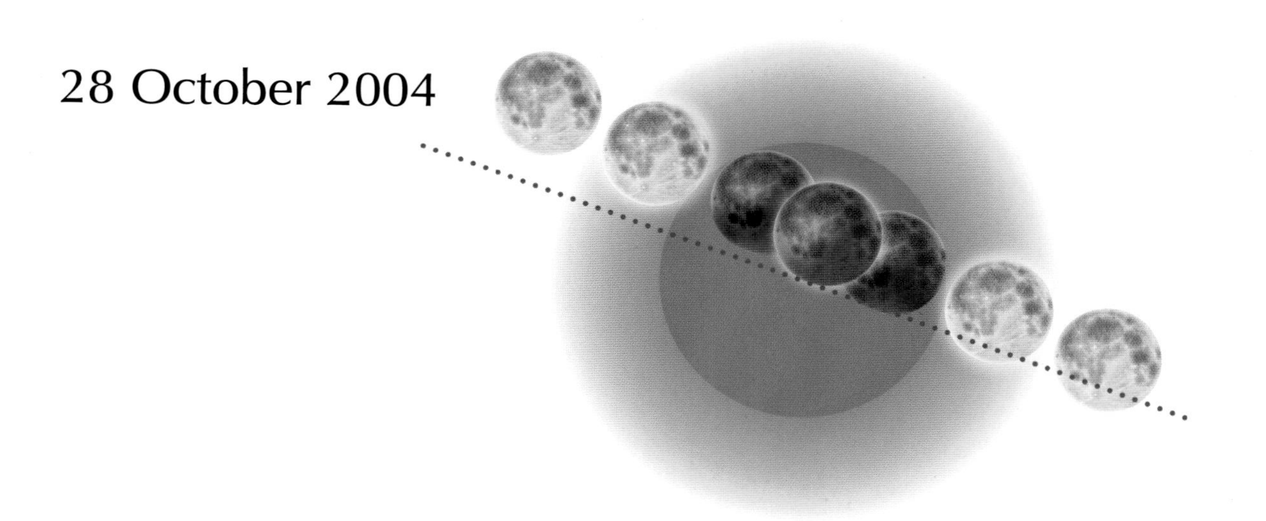
28 October 2004

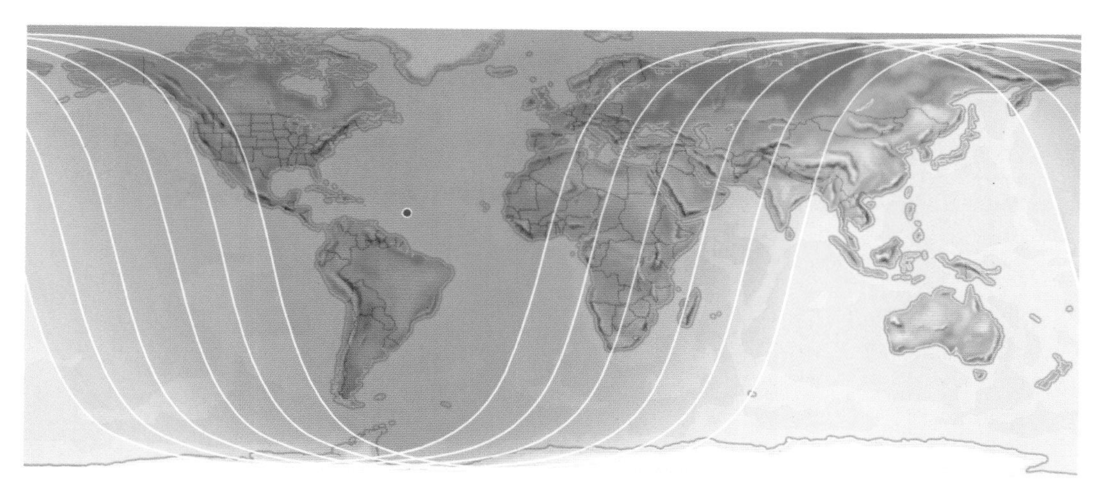

■ The eclipse of 28 October 2004 will be visible on both sides of the Atlantic, low on the eastern and western horizons. In South America, by contrast, the eclipse will be visible in the very middle of the night.

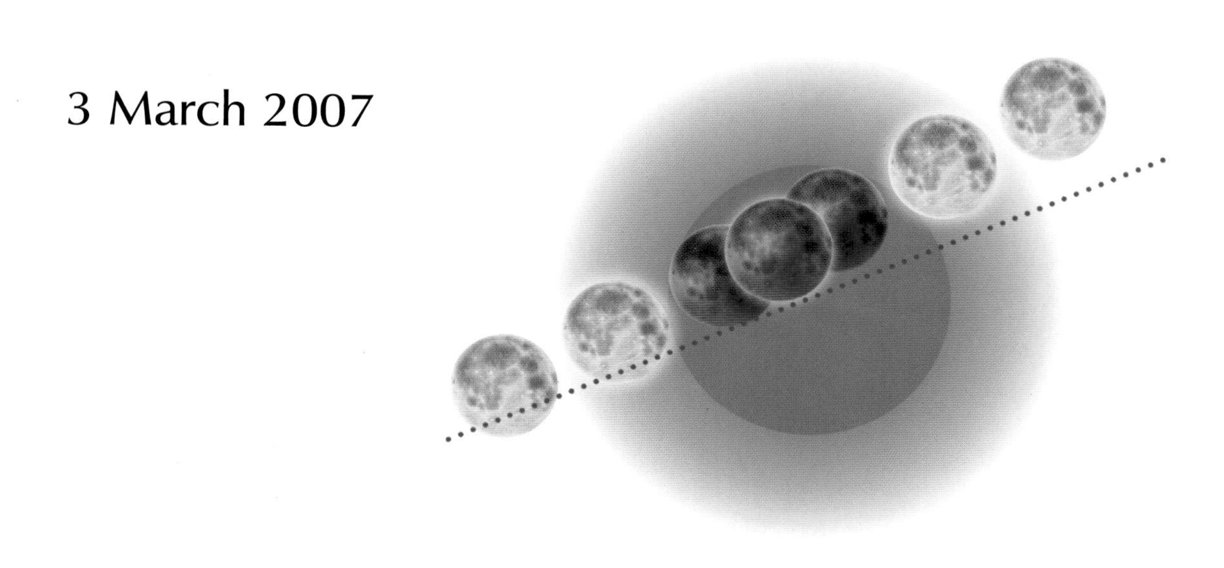
3 March 2007

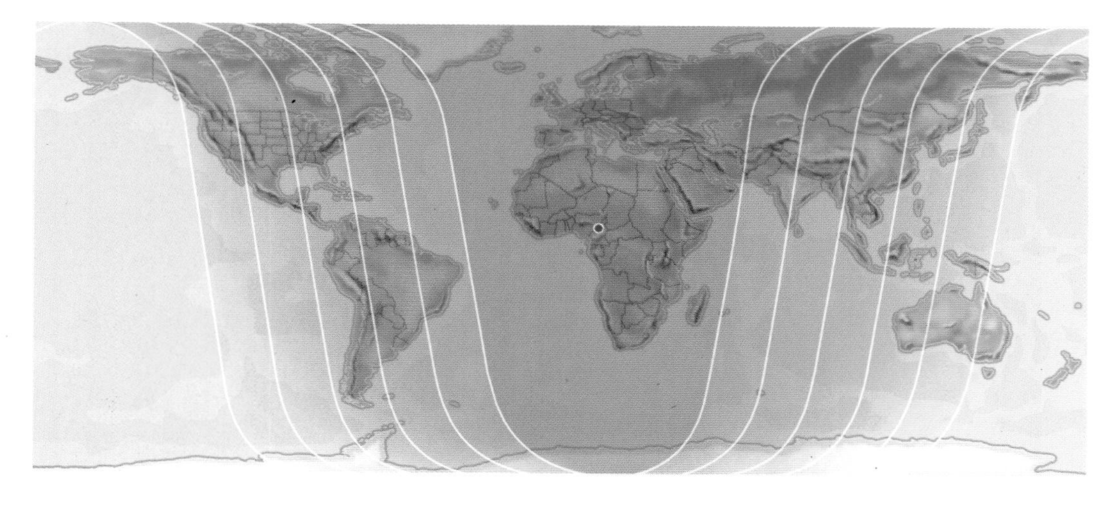

■ Eclipse chasers will not have seen a total eclipse of the Moon for nearly three years before that of 3 March 2007. The event will be visible from Africa, Europe, and part of Asia.

28 August 2007

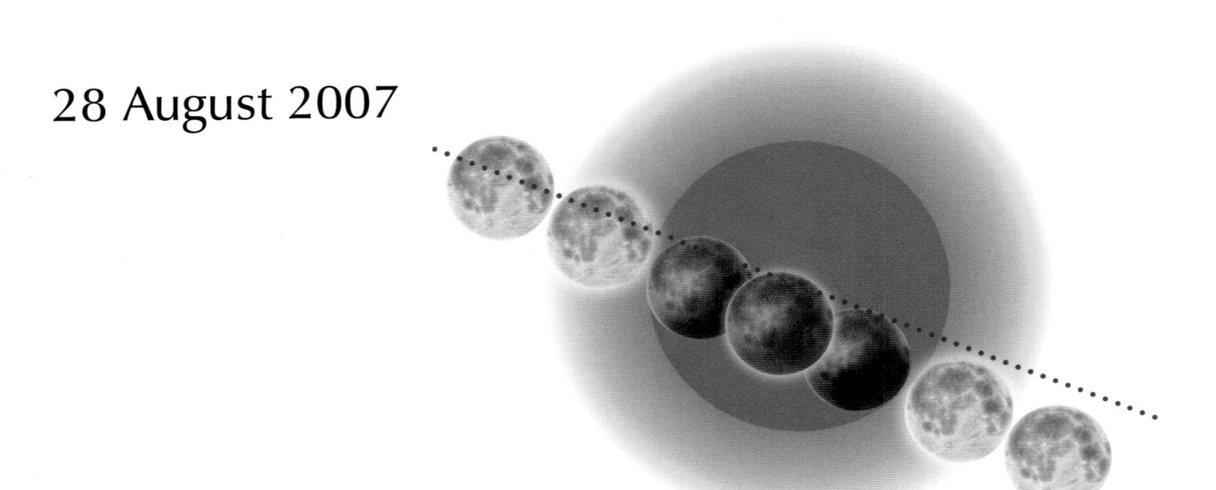

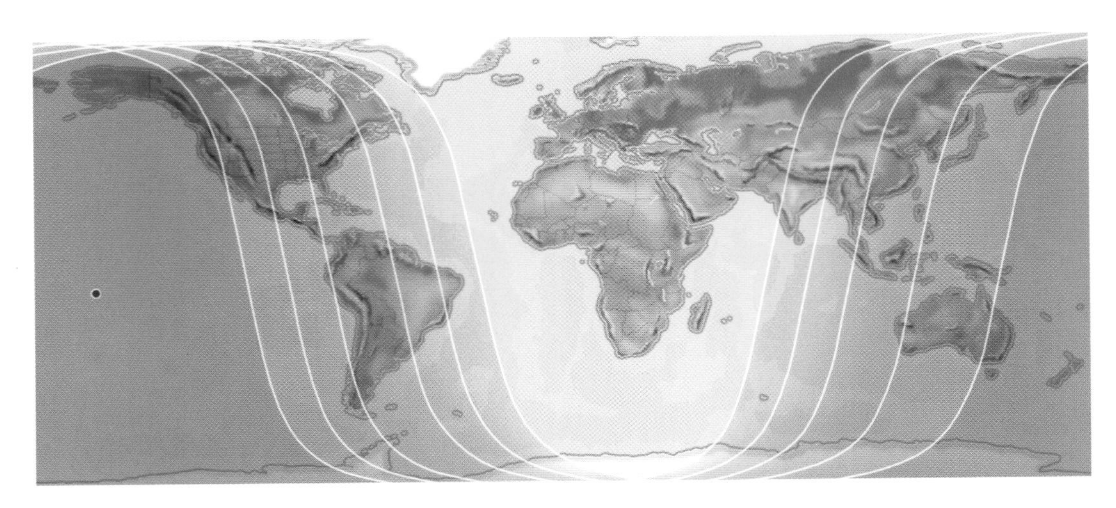

■ Exactly six lunations, i.e., nearly six months, after the total eclipse of 3 March 2007, another total occurs. Only the islands in the Pacific will see the event.

21 February 2008

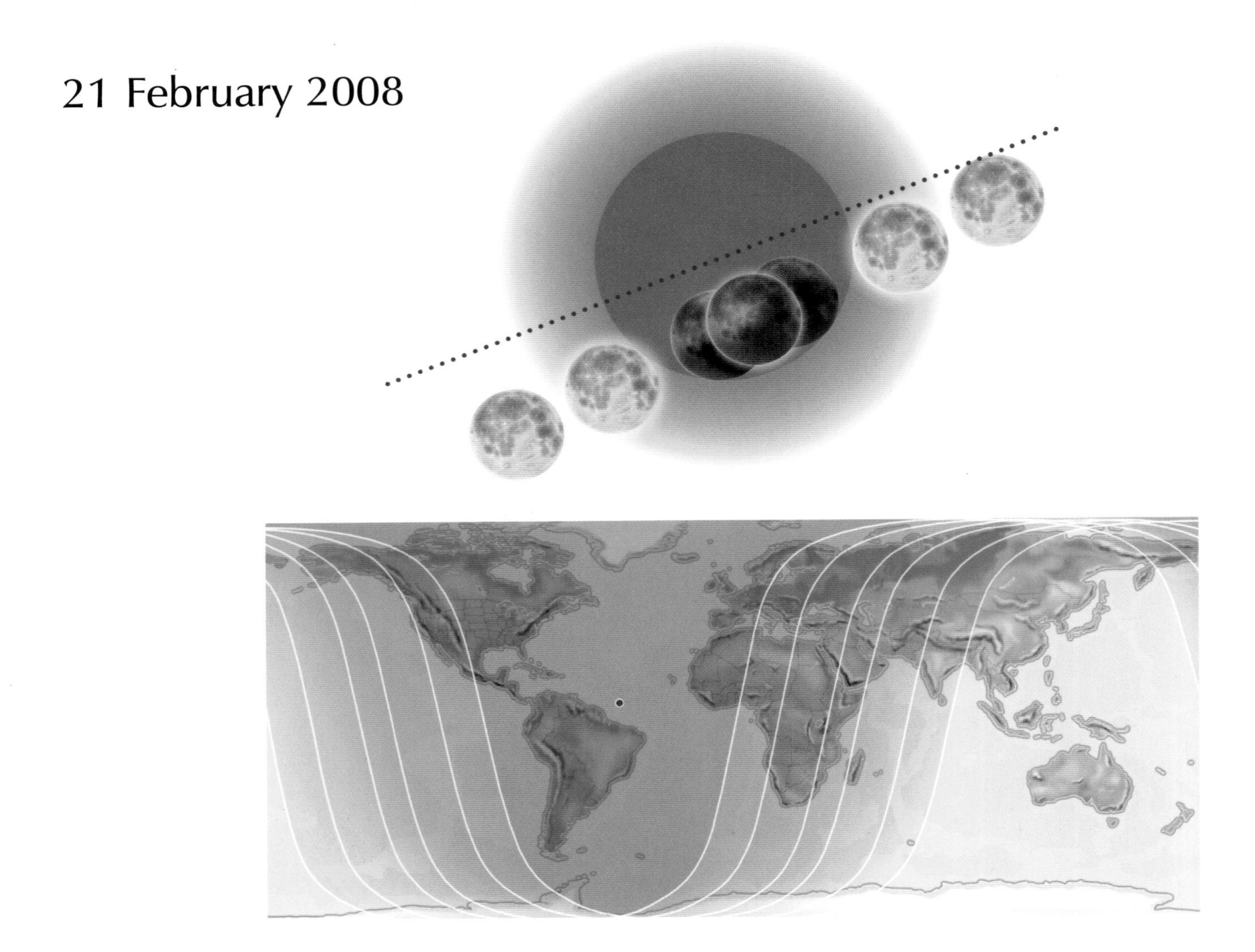

■ This eclipse is visible on both side of the Atlantic. It is impossible to predict the brightness of the Moon during the event, because it may vary by a factor of one hundred from one eclipse to another.

21 December 2010

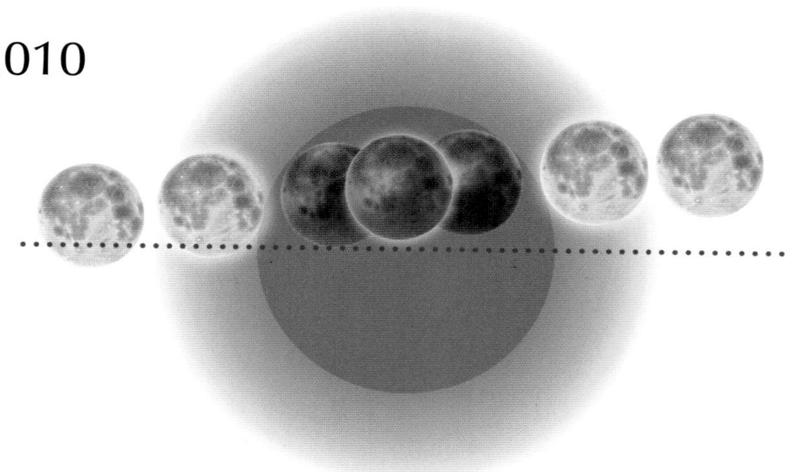

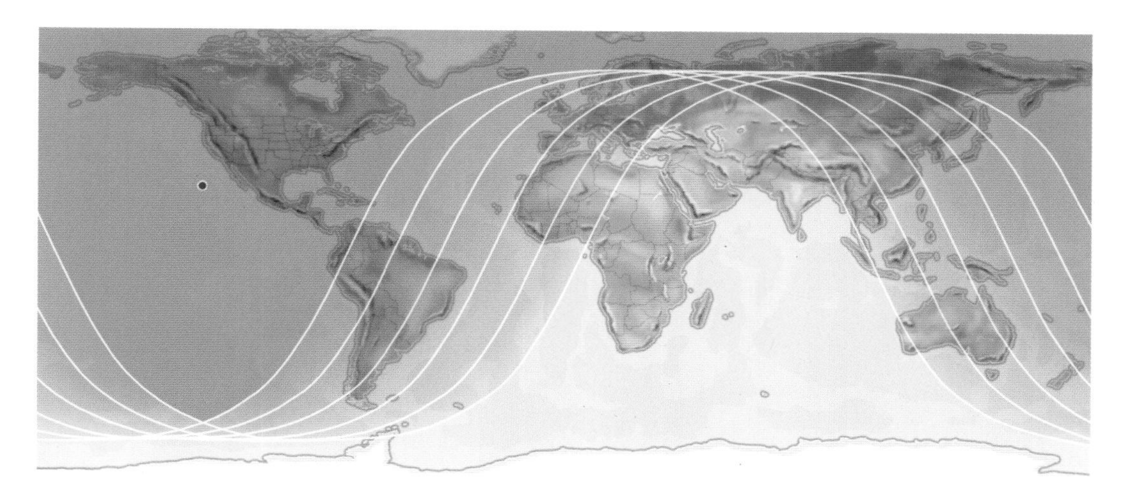

■ No less than six partial eclipses have occurred since the last total eclipse on 21 February 2008. The one on 21 December 2010 will be particularly well seen from North America.

15 June 2011

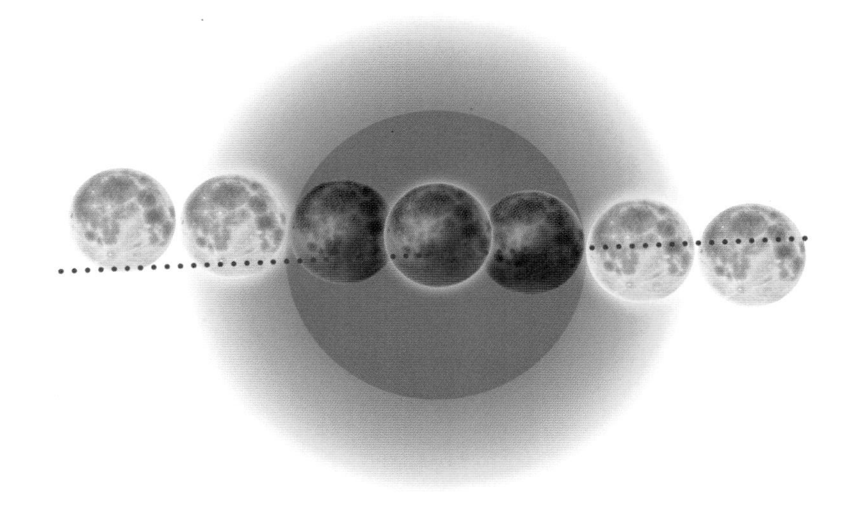

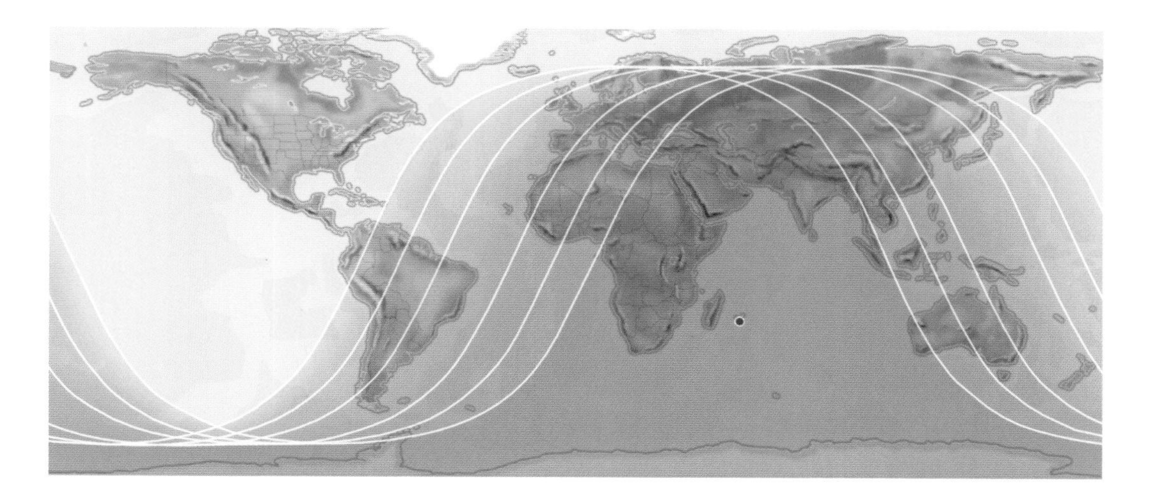

■ A fine central eclipse, perfectly visible from the Indian Ocean. The reddened Moon will be at the zenith for the island of La Réunion, and observable under excellent conditions anywhere in southern Africa.

10 December 2011

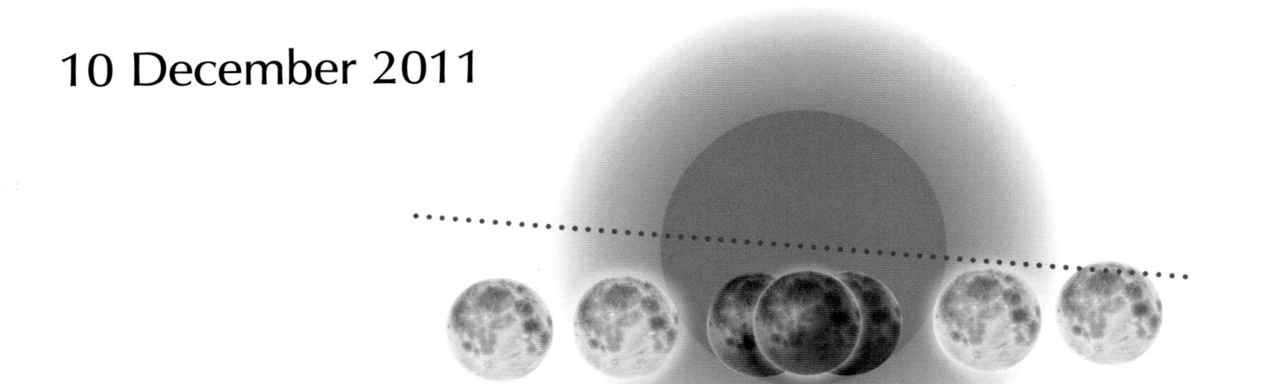

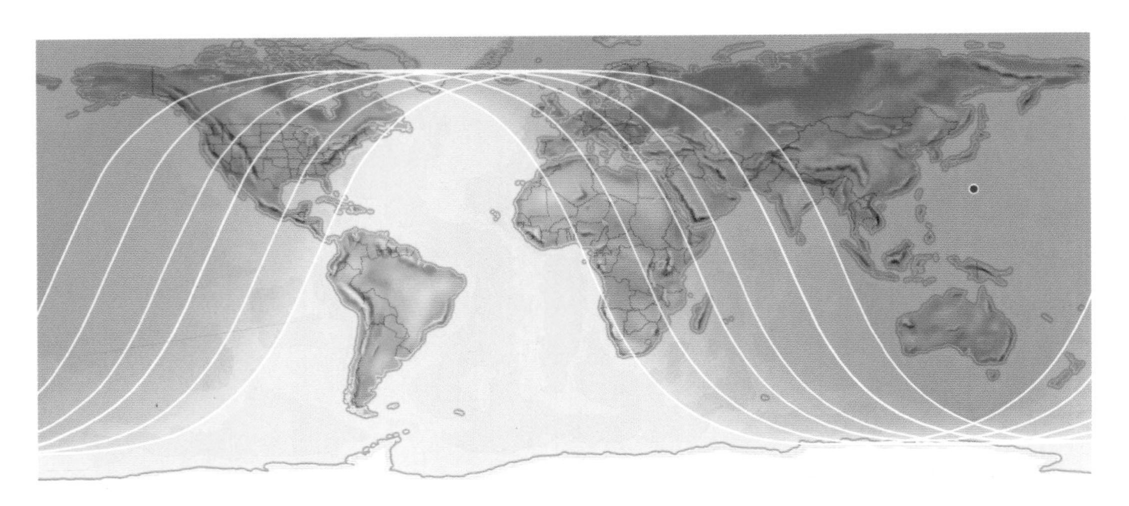

■ At the end of the southern spring, this eclipse will be of interest in Australia, Japan, and Indonesia. In Africa and Europe the eclipse is visible at sunset. In North America, it is visible at sunrise.

15 April 2014

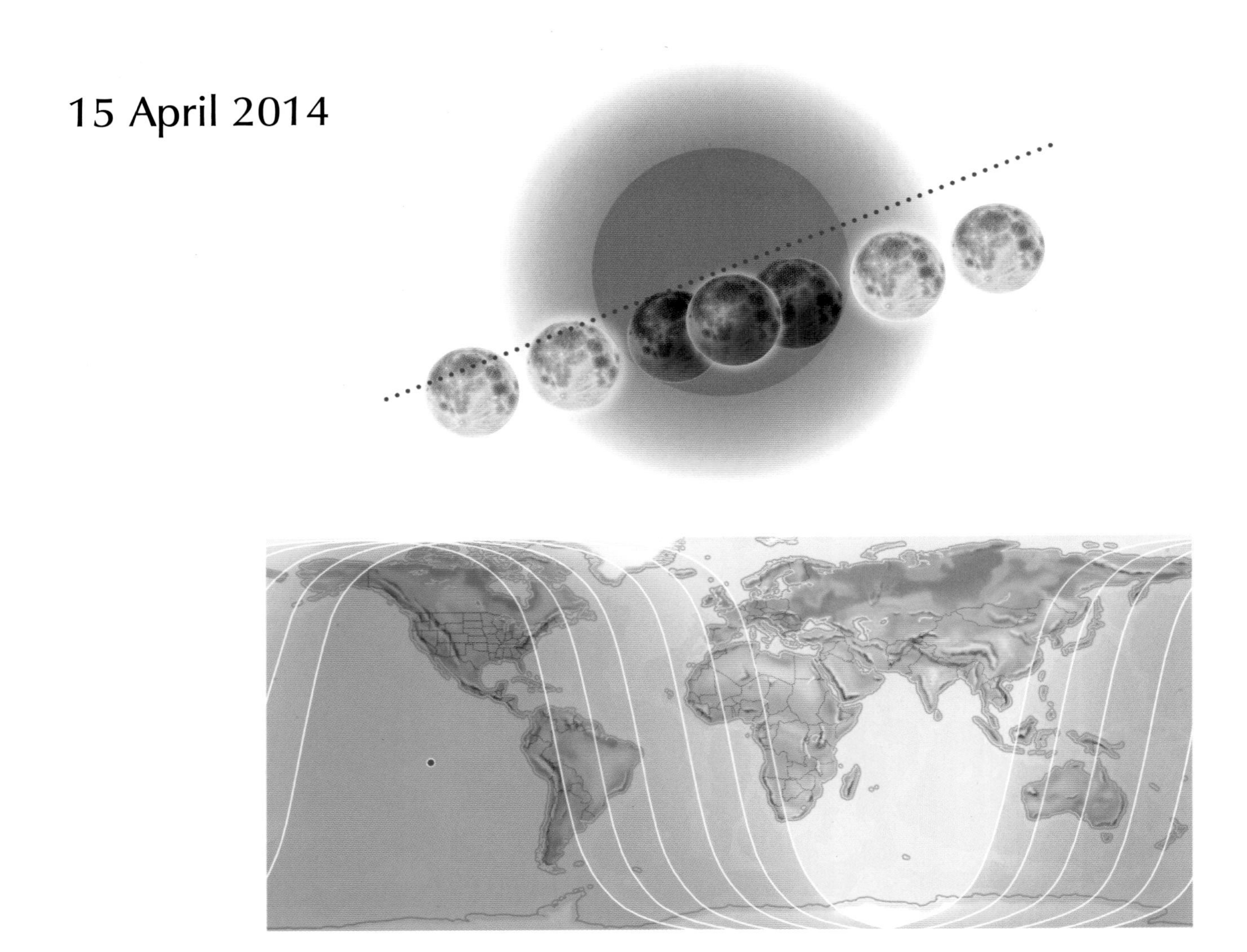

■ There will be more than two years without a total lunar eclipse, visible anywhere on the planet! On 15 April 2014, eclipse chasers will be found in the Atacama Desert, in Mexico, California, or Hawaii.

8 October 2014

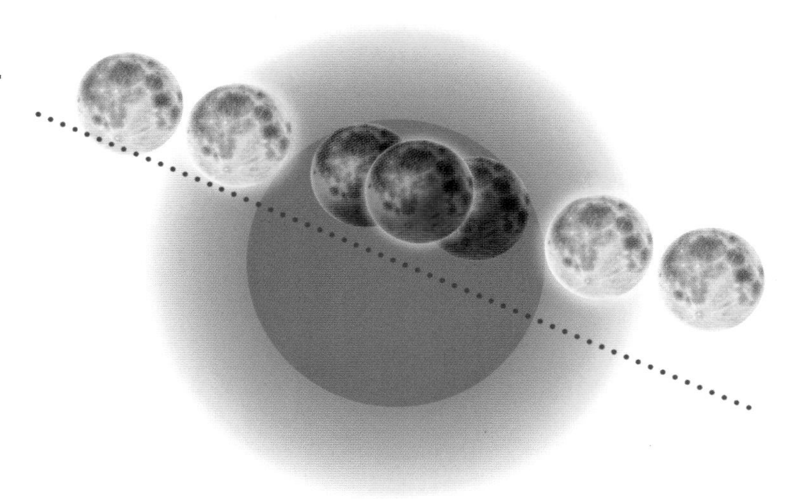

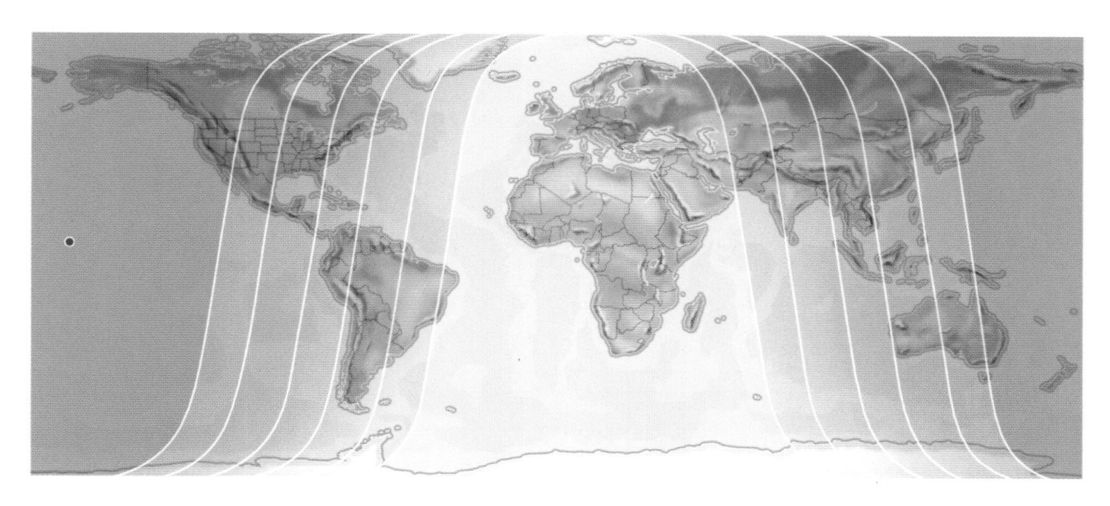

■ Six lunations after the last total eclipse, the Moon again plunges into the Earth's shadow. This time, the eclipse will be total, and almost at the zenith, above the domes of Mauna Kea Observatory, on Hawaii.

4 April 2015

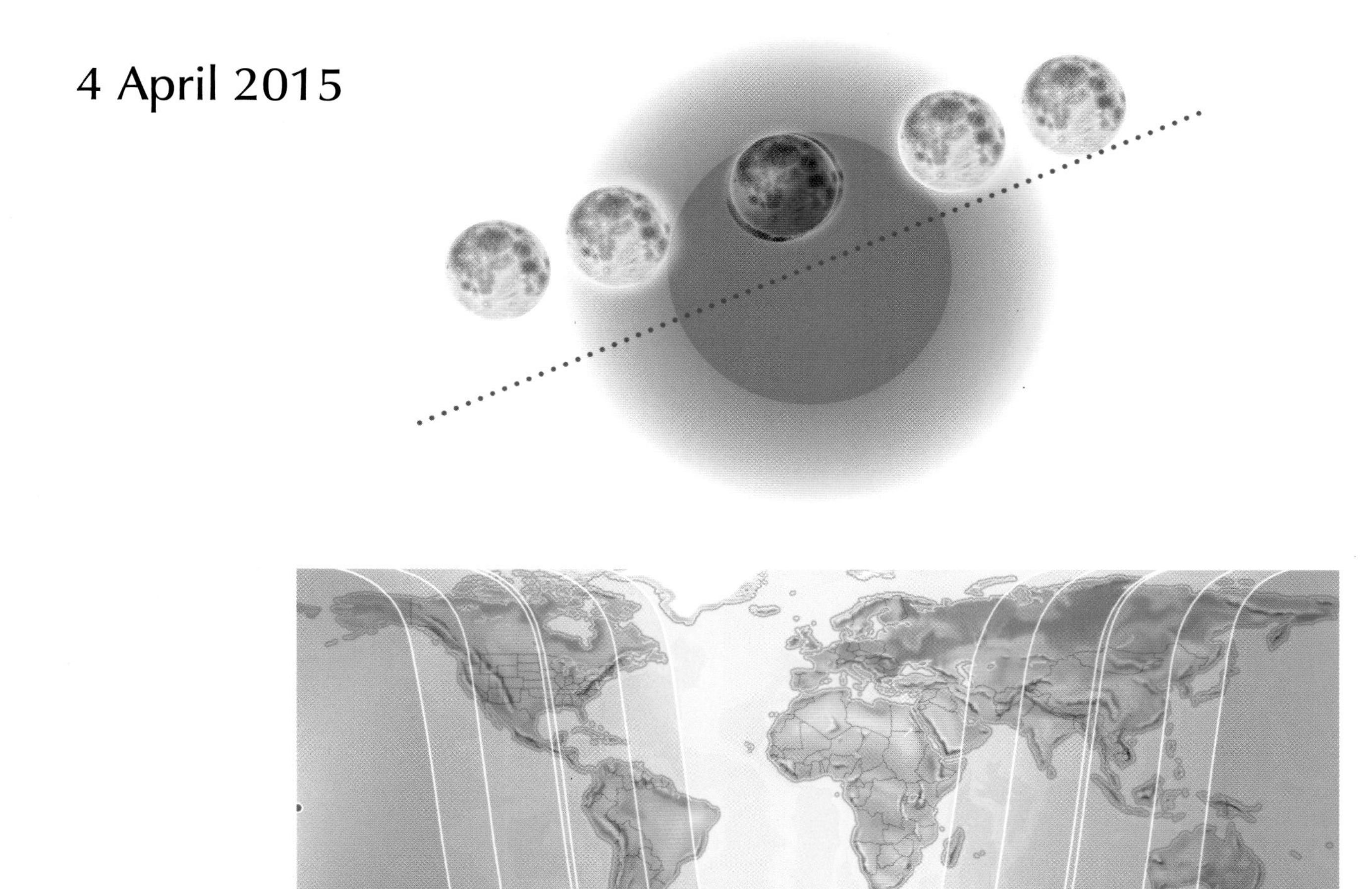

■ It is impossible to predict the Moon's appearance during the total phase in advance. The colour and brightness become apparent only when our satellite passes into the shadow of the Earth.

28 September 2015

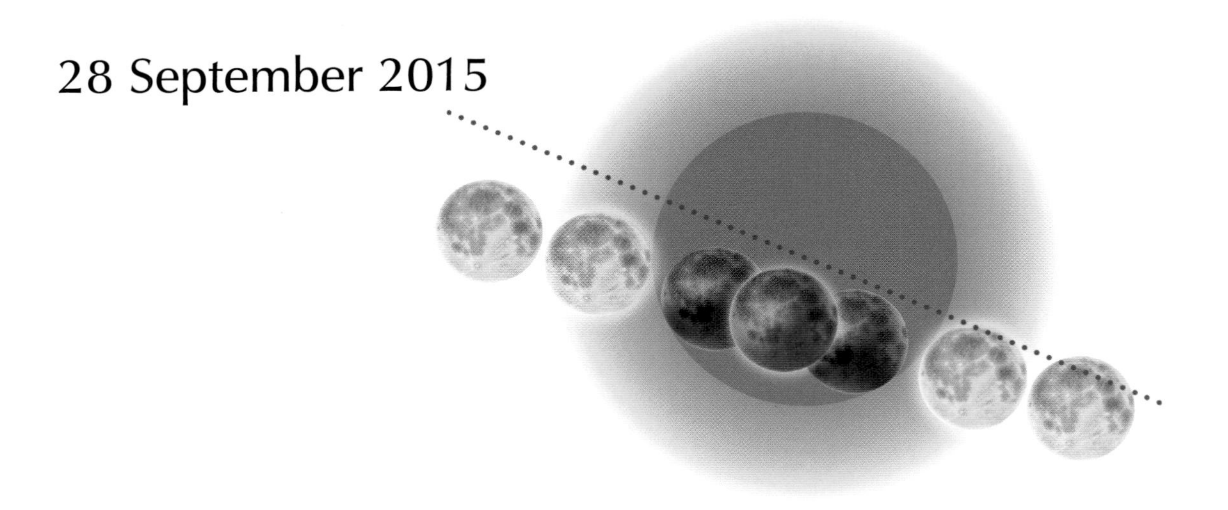

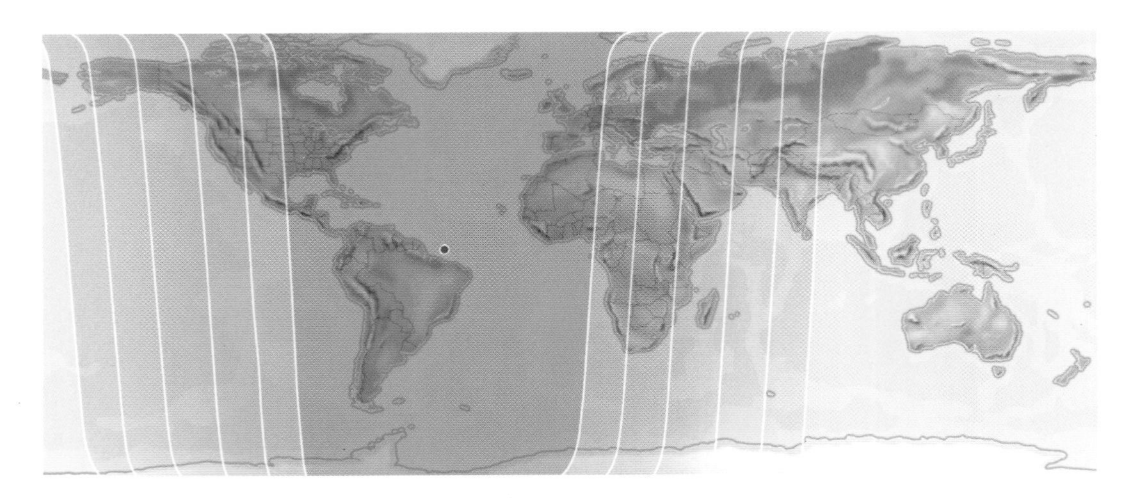

■ Eclipses of the Moon are visible from an entire hemisphere; the one where it is night-time. It is therefore possible, without making any special trips, to see several eclipses over the course of a decade.

31 January 2018

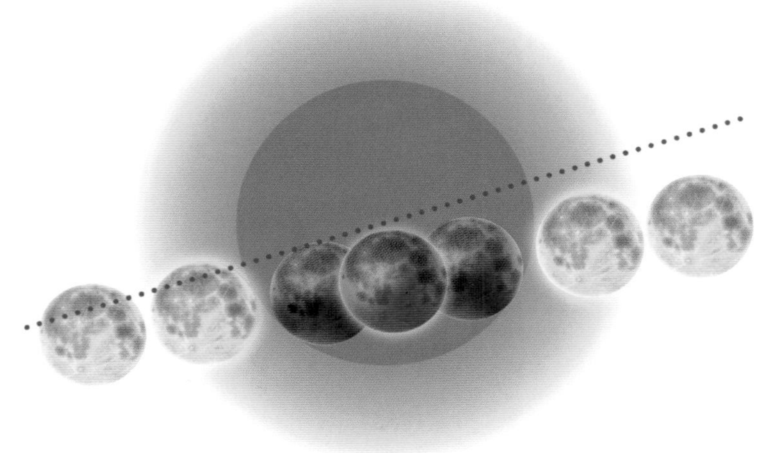

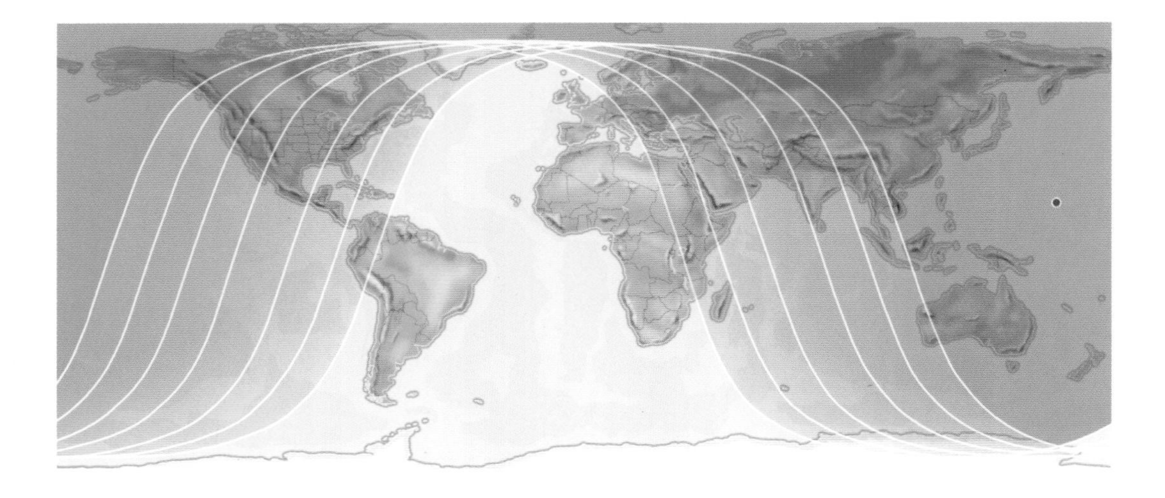

■ Five partial eclipses have occurred since the last total one, visible from Europe on 28 September 2015. On 31 January 2000, it is Australia, Hawaii, China, and Alaska that will get the best view

27 July 2018

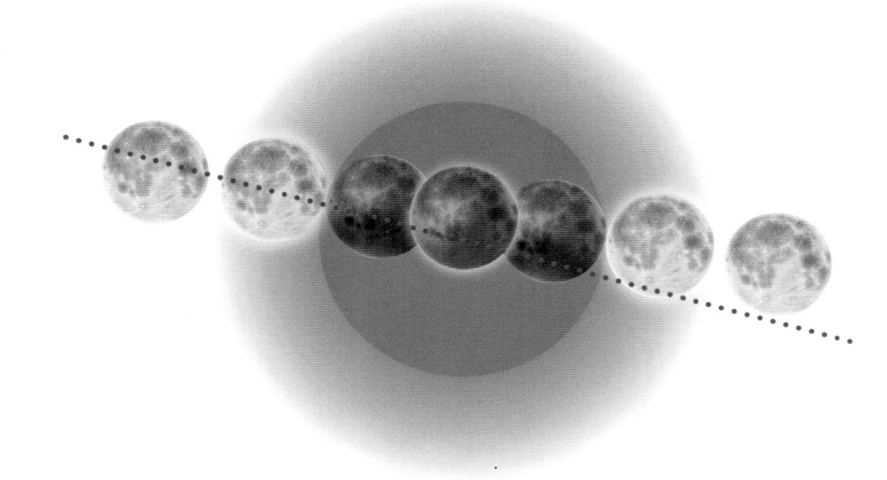

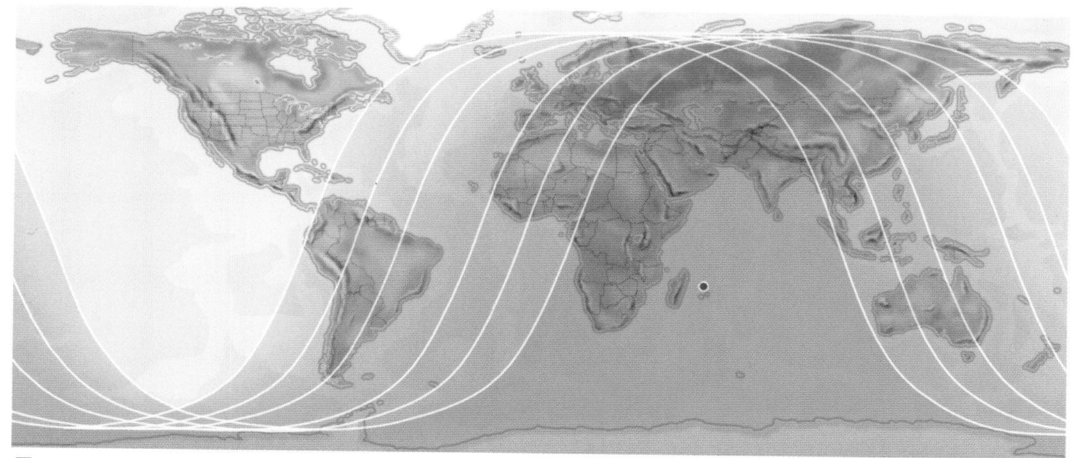

■ A splendid total and central eclipse, at the zenith for Madagascar, the island of La Réunion, and Mauritius. It is also visible from the countries bordering the Indian Ocean, at sunrise for India, and sunset for East Africa.

21 January 2019

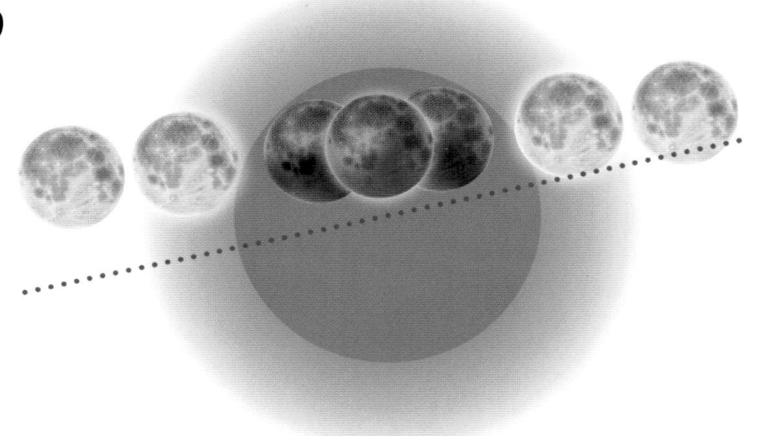

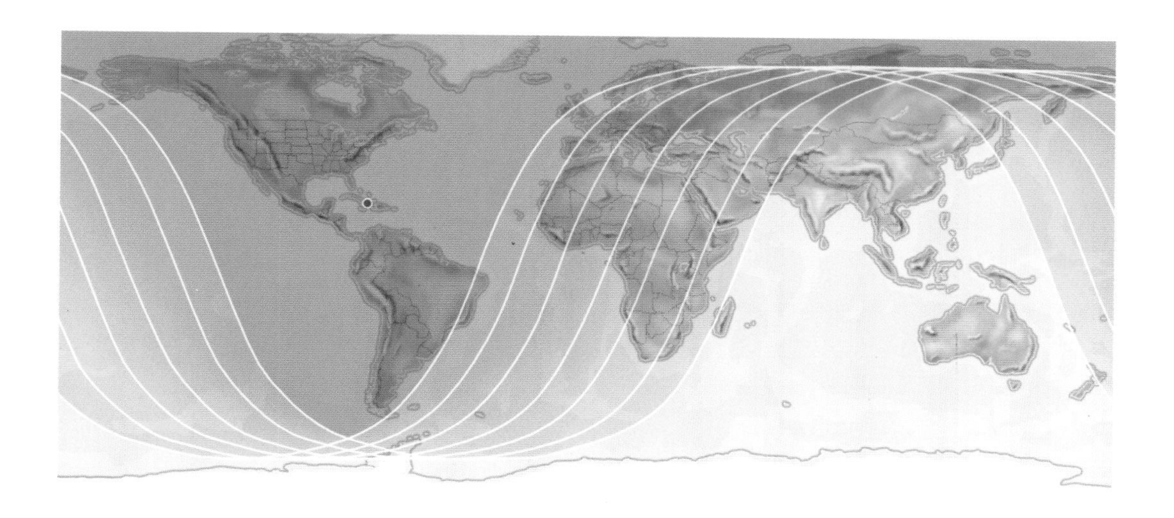

■ This is the last total lunar eclipse of the decade. Easily visible in both Americas, its ghostly light will illuminate the Alma interferometric network, in the Atacama Desert.

Glossary

ANAXAGORAS Greek philosopher (b. Clazomenes, 500 BC; d. Lampsacus, 428 BC). By assuming that the Moon (being similar to the Earth), derived its light from the Sun, he managed to explain eclipses of the Moon correctly, as occurring when, as seen from observations, the Moon passed into the Earth's shadow.

APHELION (from the Greek *apo*, far, and *helios*, Sun). The point on an orbit of a celestial body around the Sun, where its distance is a maximum. *Cf.* perihelion.

APIANUS or APIAN (Peter Bienewitz or Benniwitz, called Petrus). German mathematician, geographer and astronomer (b. Leisnig, Saxony, 1495; d. Ingolstadt, 1552). Famous for his charts and richly illustrated astronomical works, he observed Comet Halley at its return of 1531, and, on that occasion was the first in Europe, together with Fracastorius, to note that the tails of comets always point away from the Sun. He was also one of the first to propose, in his *Cosmologie* (1524), determining longitude from the motion of the Moon.

ARISTARCHUS (of Samos) Greek astronomer (Samos c. 310 BC to 230 BC). He was the first to attempt to determine, by astronomical measurements, the diameters of the Sun and Moon, and their distance relative to the Earth. Despite the inaccuracy of his measurements, he was able to establish that the Sun is distinctly larger than the Earth. He then advanced the theory that the Earth and the other planets moved around the Sun, eighteen centuries before Copernicus.

ARMILLIARY SPHERE A set of circles (or armilliaries) in metal, wood or card, that represent the principal circles on the celestial sphere, and in the centre of which is placed a small globe that represents the Earth or the Sun, and which is mounted on a stand. Formerly used as an observing instrument, for demonstration purposes, or as a purely decorative object.

BAILY'S BEADS A ring of points of light caused by dips in the profile of the lunar limb, that is observed at total eclipses of the Sun, at the beginning and end of totality.

BRAHE (Tycho) Danish astronomer (b. Knudstrup, 1546; d. Prague, 1610). Born into an aristocratic Danish family, from a very early age he was attracted to astronomy, and finally devoted himself to studying the heavens follow the appearance, in 1572, of a bright, temporary star – which was, in fact, a supernova. This amazed him, because it contradicted the theory, which had been accepted since the time of Aristotle, of the immutability of the heavens. He did, however, succeed in showing, by accurate measurements, that the object was definitely a star, at a great distance, because it showed no detectable parallax, and was not a simple meteor. This spectacular phenomenon, gave him the idea, like Hipparchus eighteen centuries earlier, of compiling an accurate catalogue of stars. He was able to realise this ambition, thanks to the liberality of the King of Denmark, Frederick II, who, in 1576, presented him with the island of Hveen, in the Sound, not far from Copenhagen, and also gave him a pension. Tycho wasted no time in having a great observatory Uraniborg ('palace of Urania') built in the centre of the island, and eqipping it with the largest and most accurate instruments of the time. Then, in 1584, he set up a second observatory, Stellenborg ('palace of the stars') close to the earlier one. Here the instruments, protected by movable domes, were installed in the basement so that they would be sheltered form the violent winds that frequently swept the island. In addition to his two observatories, which foreshadowed great modern observatories, Tycho Brahe had workshops for maintaining his instruments, and paper-making, as well as a printing works for publishing the results of his research and of his numerous students. For more than twenty years, until 1597, Tycho Brahe accumulated thousands of observations from Hveen. Thanks to his work, astronomical observations attained the highest degree of accuracy possible before the invention of the telescope. For the first time, angular measurements reached an accuracy of about 0.3′, and were corrected for atmospheric refraction. Among his discoveries were: the variation in the obliquity of the ecliptic; two inequalities in the motion of the Moon; variation and the annual equation; the variation in the inclination of the lunar orbit relative to the ecliptic, etc. In 1577, his observations of a bright comet enabled him to establish that the latter was not an atmospheric phenomenon, contrary to the view accepted since Aristotle, but was a body lying well beyond the Moon. Finally, he left a catalogue of 777 stars. His accurate observations of the motion of the planet Mars allowed Kepler, after his death, to deduce the laws of planetary motion. A brilliant observer, Tycho Brahe was less inspired as a theoretician. Challenging the heliocentric universe for both physical and metaphysical reasons, he rejected Copernicus' system, but not by accepting Ptolemy's scheme, whose deficiencies he recognized. He proposed a hybrid system in which the planets revolved around the Sun, with the latter (and its planetary attendants) revolving around the Earth. He also proved to be an extremely bad administrator. At the death of his protector, Frederick II (1588), his pension was removed, and he was disgraced. In 1597 he had to leave Denmark and, in 1599, he took refuge in Prague, as astronomer to the Emperor of Austria, Rudolph II. Stubborn and an individualist, throughout his life he often showed an arrogant nature. As a young man he fought in a duel, when his opponent sliced off his nose with his sword. Tycho Brahe then had an artificial one made of gold and silver, which he wore for the rest of his life.

■ In King Solomon's Mines, Allan Quatermain gains the confidence of the Kukuana – guardians of the diamond mines – by convincing them that an eclipse of the Moon is a sign indicating that his servant is their king.

CALENDAR A system of ordering and counting days and years for the purposes of civil or religious life, and also for chronology.

CHROMOSPHERE A layer in a star's atmosphere, particularly of the Sun, that lies between the photosphere and the corona.

CLAIRAUT (Alexis) French mathematician (Paris 1713–1765). A precocious child, he presented his first paper to the Acadèmie des Sciences at the age of 12, and was admitted as a member at the age of 18, after having been granted an age dispensation by the king. In 1736, he was sent to Lapland with Maupertuis to determine the length of a degree of the meridian. Shortly after his return, he published his *Theory of the shape of the Earth* (1743), which did much to cause Newton's theory of gravitation to be accepted in France. His work, *Theory of the Moon* (1752) proposed an initial approximate solution to the three-body problem – the motion of three material bodies that attract one another according to Newton's laws. He applied his result to the movement of Comet Halley and, taking account of the perturbations on the comet's orbit caused by the presence of Saturn and Jupiter, he calculated, to about half a month, the date of perihelion passage, with useful assistance from Lalande and Mme Lepaute.

CONJUNCTION The situation when two or more bodies have the same geocentric longitude or right ascension.

CORONA The outermost region of the atmosphere of a star, in particular of the Sun. It is not at all homogeneous, very tenuous, and fades gradually into space.

CORONAL STREAMER A narrow, bright, and approximately radial structure in the solar corona.

CORONOGRAPH A special astronomical refractor, which enables the solar corona to be studied or photographed outside the time of total solar eclipses.

CURVATURE OF SPACE According to the theory of relativity, a geometrical property of space-time, whose most readily detectable effect is gravitation.

DRACONITIC (from the Greek *drakôn, -ontos*, dragon) Applied to the periodicity of motion in a moving reference frame, in which the fundamental plane is the ecliptic, and the origin is the ascending node of the lunar orbit. Draconitic month: average value of the interval of time between two consecutive passages of the Moon through its ascending node.

EINSTEIN (Albert) American physicist of German origin (b. Ulm, 1879; d. Princeton, New Jersey, 1955). After secondary studies at Munich that were not noted for their brilliance, he entered the Federal Institute of Technology in Zurich. In 1902 he found a job at the Federal Patent Office in Bern. Appointed, not without some difficulty, to head of courses at Zurich University in 1909, he remained there for just two years. After a stay at Prague University (1911–12), followed by a brief return to Zurich, he accepted, after mature consideration, a post at the Kaiser-Wilhem Institute in Berlin. He remained there until Hitler came to power, without becoming completely established, because he travelled extensively and spent a lot of time at foreign universities. Forced to leave Germany in 1933, he first settled in Paris, then in Belgium, before accepting the first professorial chair at the Institute for Advanced Studies at Princeton, where he remained until the end of his career. By demonstrating the existence of the photon, establishing the equivalence between mass and energy, and by developing the theory of relativity (in its Special form in 1905, and generalized in 1916), which revised all notions of space and time, and introduced a new concept of gravitation, he revolutionized physics and astronomy, and reinvigorated cosmology.

EXPANSION (of the universe) A phenomenon first envisaged as arising from the theory of General Relativity by W. De Sitter, in 1917, and then by A. Friedmann in 1922 and G. Lemaître in 1927, and in which the different galaxies in the universe are receding from one another at a velocity that is proportional to their respective distances.

FLAMMARION (Camille) French astronomer (b. Montigny-le-Roi, Haute-Marne, 1842; d. Juvisy-sur-Orge, 1925). He was an enthusiastic and talented popularizer of astronomical knowledge of the period. His book, *Astronomie populaire* (1879) had an immense success, and prompted many to become astronomers. In 1887, he created the Société Astronomique de France. Settling at Juvisy, south of Paris, he installed an observatory there in 1883, where, until his death, he carried out a wide range of research (astronomy, meteorology, climatology), obtaining, in particular, numerous observations of the planets.

GALILEI (Galileo) Italian astronomer and physicist (b. Pisa, 1564; d. Arcetri, 1642). He is one of the founders of modern mechanics and played a major role in introducing mathematics to explain physical laws. In astronomy, he introduced the use of the telescope, which caused a revolution in techniques for the observation of the universe. Thanks to the rudimentary refractors that he constructed from 1609 onwards, and which he had the idea of turning towards the sky, he made the first observations of relief on the Moon, discovered the phases of Venus and the four principal satellites of Jupiter, understood that the Milky Way consisted of a vast assembly of stars, and was able to detect more stars than had ever been previously suspected. These discoveries, which were announced in 1610 in the *Siderius nuncius (Messenger of the Stars)*, clearly demonstrated that the universe did not agree with the characteristics descibed by Aristotle, and came to reinforce the ideas of Copernicus in various ways. The presence of mountains on the Moon (whose heights Galileo found by measuring the lengths of their shadows) and of valleys, proved that it was not fundamentally different from the Earth and, because it moves across the sky, it was no longer absurd to imagine that the Earth was also moving. The appearance of the phases of Venus, which was inexplicable in Ptolemy's system, offered indisputable empirical proof of the heliocentric system, the validity of which was confirmed by the presence of satellites orbiting Jupiter, forming a miniature Solar System in themselves. Finally, the distinction between stars and planets that was provided by the telescope – the former remaining points of light, whereas the latter showed a distinct apparent diameter – proved that stars are considerably farther away than the planets, thus revealing the immense size of the stellar universe.

HALLEY (Edmond) English astronomer and physicist (b. Haggerston, London, 1656; d. Greenwich, 1742). The author of numerous works concerning geophysics, meteorology, and astronomy, he was a colleague of Newton's, whom he persuaded to publish the Principia. He remains primarily known for his study of the motion of comets. In his Synopsis of the Astronomy of Comets (1705), he identified the comets that appeared in 1531 and 1607 with the one that he had observed himself in 1682, thus demonstrating that it was a periodic comet, whose return he predicted for 1758. From 1720 to his death, he was Astronomer Royal and head of Greenwich Observatory, where he succeded J. Flamsteed.

HIPPARCHUS Greek astronomer (2nd century BC). He may be considered the founder of positional astronomy. His extensive work, known to us thanks to Ptolemy, includes the accurate measurement of the orbital period of the Moon, the determination of the Earth–Moon distance, establishment of tables of motion of the Sun and Moon, etc. His principal discovery was that of the precession of the equinoxes: having noted that the Sun, in the course of its apparent annual motion takes slightly more time to return to the same point in the Zodiac than it does to reach the celestial equator between one spring and the next, he correctly explained the phenomenon by a slow displacement, relative to the stars, of the equinoctial points, i.e., the points where the ecliptic and the celestial equator intersect. In 134 BC, the sudden appearance of a nova prompted him to measure accurately the celestial coordinates of stars so that it would be possible to detect any new, unexpected apparition, and to reveal any possible motion of the stars relative to one another. He compiled a catalogue of 1025 stars, where they were classified, for the first time, in six magnitudes depending on their brightness. He also laid the basis for trigonometry, invented stereographic projection (used for the construction of astrolabes), and suggested the first method of determing longitude.

JANSSEN (Jules) French astronomer (b. Paris, 1824; d. Meudon, 1907). His scientific career started with optics. In 1862, he installed a small personal observatory in the roof of his house in Montmartre, and started research into solar spectroscopy. He was able to establish the terrestrial origin of certain lines in the solar spectrum, which he proposed calling, for that reason, telluric lines. While still pursuing the study of these lines, he subsequently undertook the spectroscopic study of planetary atmospheres and discovered, in 1867, the presence of water vapour in the atmosphere of Mars. In 1868, he travelled to India to observe a total eclipse of the Sun. On this occasion, at the same time as Lockyer, he discovered the presence in the solar atmosphere of a chemical element, helium, then unknown on Earth, through spectral analysis of prominences. In 1874, he went to Japan to observed the transit of Venus using a new technique, chronophotography, which foreshadowed cinematography. In 1874, it was decided to create an astrophysical observatory near Paris, and Janssen had to decide between two sites: la Malmaison, or the old chateau at Meudon. He chose Meudon and, in 1876, started to develop the new observatory, which he directed until his death. In particular, he installed a large refractor with an objective 83 cm in diameter, and a photographic objective of 62 cm, together with a 1 m telescope. He made the new establishment into an important centre for solar and planetary astrophysical research. Between 1887 and 1897, he made several expeditions to the top of Mont Blanc to take advantage of the sky's transparency at high altitude. He even built an observatory there, which required the transport of 15 tonnes of equipment by manpower, but which was engulfed by the snow after several years. Finally, he was one of the pioneers of astronomical and meteorological observation from a balloon.

KEPLER (Johannes) German astronomer (b. Weil – now Weil der Stadt – Wurtemberg, 1571; d. Ratisbon, 1630). He is one of the founders of modern astronomy. Of humble origins, he was given a free scholarship to the seminaries of Adelberg (1584) and Tübingen (1589), where one of the most ardent defenders of the Copernican theory, Mästlin, introduced him to astronomy. Professor of mathematics at Graz, he was driven out around 1600 by religious persecution. He took refuge in Prague, where he became the disciple and assistant to Tycho Brahe, to whom he succeeded in 1601, as astronomer to Emperor Rudolph II, and then to Emperor Mathias, who nominated him as professor of mathematics at Linz. At the insistence of the Duke of Wallenstein, he moved to Ulm. Short of money, he was reduced to living by casting horoscopes and selling small almanacs. A partisan of the heliocentric system, he explained, in an early work, the *Prodomus ... mysterium cosmographicum*, published in 1596, why the Ptolomaic system needed to give way to the Copernican representation of the world. But, haunted by Pythagorean ideas, he believed that the universe was constructed on the basis of a geometrical architecture. He therefore developed an ingenious geometrical model of the Copernican system in which the orbit of each planet occupied a sphere that circumscribed a regular polyhedron and was inscribed within another. In fact, Kepler was convinced that the number of planets, their distances from the Sun, and the orbital velocities were not the result of chance. He gave himself the goal of finding the laws of their motion, as well as those governing the distribution and size of their orbits. It was in devoting himself to a systematic study of the motion of Mars (whose movement remained poorly described by both the Ptolomaic and Copernican systems), and after laborious calculations, which he checked with Tycho Brahe's accurate observations, that he discovered the first two of the laws that have imortalized his name. These were published in 1609 in his Astronomia nova. Still plagued by Pythagorean ideas, he then struggled to show the existence of a harmonic relationship (in the musical sense of the term) between the highest and the lowest velocity of the planets. He thus discovered the third fundamental lay of planetary motion, which established a relationship between the size of planetary orbits and the time required to complete them. He published this law in 1619 in his Harmonices mundi, where he described his somewhat mystical vision of the universe. In the last years of his life he devoted his time to establishing tables of the positions that were as accurate as possible, based on the laws that he had discovered and on Tycho Brahe's observations: the Rudolphine Tables, which he published in 1627.

LUNATION The period of time between two consecutive New Moons, the average duration of which is 29 d 12 h 44 m 2.9 s, and were fluctuations may amount to 13 h on either side of this value. Syn: lunar month.

LYOT (Bernard) French astronomer (b. Paris 1897; d. Cairo, 1952). Assistant to Pérot at the École polytechnique, he joined Meudon Observatory in 1920. He concentrated on the study of the polarization of the light scattered by the Moon and planets, and to this end, developed a polarimeter with a high sensitivity. Subsequently, interested in the outer layers of the Sun, he invented and constructed the coronograph in 1931, which enabled observation of the corona outside total eclipses. Thanks to this, in 1935 he obtained the first film showing the motion of prominences. In 1933, he invented a monochromatic polarizing filter, which selected the appropriate radiation for viewing different structures on the Sun, and gave the first monochromatic images. In 1939 he was elected to the Académie des Sciences. In 1948, thanks to the progress of electronics, he developed the photoelectric photometer, the principles of which he had described in 1924. He died returning from a mission to Khartoum to observed a total eclipse of the Sun. He was one of the people who, before the advent of the space age, made the greatest contribution to our knowledge of the Sun and planetary surfaces.

MAUNDER (Edward Walter) British astronomer (b. London, 1851; d. London, 1928). After joining Greenwich Observatory in 1873 – and where he passed his whole career – he set up a department that specialized in the daily photographic observation of the Sun. Primarily interested in sunspots, by examining earlier observations, he famously established that the Sun was completely devoid of spots between 1645 and 1715 (which is now known as the Maunder Minimum). With his wife Annie Russell and his brother Thomas, he did a lot for the development of popular astronomy in Great Britain. He was one of the founder members of the British Astronomical Association (1890).

METONIC CYCLE A period of 235 lunations found in the 5th century BC by the Athenian astronomer Meton, and adopted in Greece in 432 BC to ensure that the solar and lunar years were in agreement.

NEW MOON Lunar phase corresponding to the beginning of a lunation, when the Moon, for a terrestrial observer, is practically in the same direction as the Sun. (The dark hemisphere is then facing the Earth, so it is not observable in the sky.)

NODE Each of the intersections of the orbit of a body, orbiting a central body, with a reference plane.

OCCULTATION Temporary disappearance of a body behind another with a greater apparent diameter.

OPPOSITION The configuration of two bodies, whose angular separation on the celestial sphere is 180°.

ORBIT The path followed by a body around another or in a system of several bodies, under the influence of gravitation.

PENUMBRA 1. Zone surrounding the shadow cone of a body in the Solar System, and where only part of the solar disk is visible. (During the course of total or partial lunar eclipses, the penumbra surrounds the shadow cast by the Earth on the Moon.) 2. Outer region of a sunspot, which has an approximately radial filamentary structure.

PERIHELION The point on the orbit of a celestial body at which it is closest to the Sun. Cf. aphelion. Advance of perihelion: the slow rotation of the orbit of a planet in its own plane, and in the same sense at the orbital motion, and expressed as a secular term in the longitude of perihelion. (Classical celestial mechanics explains most of this advance, leaving, particularly in the case of Mercury, a residual, which the theory of General Relativity is able to explain.)

PHASE (from the Greek *phainen*, shine, appear). The successive stages in the appearance of the Moon and planets in the course of a synodic period, and which depends on their position in space relative to the Sun and Earth. (The Moon's cycle of phases consititutes the lunation, or lunar month. Among the planets, only Mercury and Venus, which are closer to the Sun than the Earth, show marked phases for a terrestrial observer.) Phase angle: the angle at the centre of the illuminated object between the direction of the illuminating body (the Sun for an object in the Solar System, or primary star in an eclipsing binary), and that of the observer.

PROMINENCE (solar) A high, narrow structure in the solar atmosphere.

RELATIVITY (theories of) A set of theories that maintain that are equivalent reference frames for the description of phenomena, with the values relative to one reference frame being deduced, according to certain transformations (proper to each theory), from the same relative values in another reference frame, with the physical laws that express the relationships between these values remaining invariant.

SAROS A period of 18 years 11 days (or 18 years 10 days, if there are 5 leap years in the interval) which approximately governs the return of solar and lunar eclipses.

SOLAR ACTIVITY Various phenomena related to perturbations in the Sun's magnetic field, of varying duration and intensity.

SOLSTICE (from the Latin *sol*, Sun, and *status*, from *stare*, stand still). 1. A time of year when the Sun, in its apparent motion along the ecliptic, attains its greatest northern or southern declination, and which corresponds to a maximum or minimum length of day. 2. The corresponding point on the Sun's apparent path along the ecliptic.

SPECTRUM 1. The range of monochromatic radiation obtained by dispersing light or, more generally, any complex radiation. 2. The range of radiation emitted, absorbed, scattered, etc. by an element or chemical species, under specific conditions.

SUNSPOT A dark, temporary, evolving feature, which has an approximately circular shape of variable size and is found on the Sun's photosphere.

TIDE 1. Daily oscillation of the oceans, the level of which alternately rises and falls as a result of the gravitational attraction of the Moon and the Sun. 2. Any deformation of a body caused by the gravitational attraction of another.

Index

Bibliography

Audouze, J. & Israël, G., *The Cambridge Atlas of Astronomy,* 3rd edn., Cambridge Univ. Press, Cambridge, 1994

Brewer, B., *Eclipse*, Earth View, Seattle, 1991

Dreyer, J.L., *History of Astronomy from Thales to Kepler*, Dover, New York, 1953

Espenak, F., *Fifty Year Canon of Lunar Eclipses: 1986–2035*, Sky Publishing, Cambridge, Mass., 1989

Espenak, F., *Fifty Year Canon of Solar Eclipses: 1986–2035*, Sky Publishing, Cambridge, Mass., 1987

Espenak, F., Internet site: http://sunearth.gsfc.nasa.gov/eclipse/eclipse.html

Flammarion, C., *Popular Astronomy*,

Gore, J.E., Chatto & Windus, London, 1884

Harrington, P.S., *Eclipse!*, John Wiley and Sons, 1997

Harris, J. & Talcott, R., *Chasing the Shadow*, Kalmbach Books, Milwaukee, 1994

Illingworth, V. (ed.), *Collins Dictionary of Astronomy*, HarperCollins, London, 1994

Koutchmy, S. & Guillermier, P., *Total Eclipses*, Springer-Verlag, Heidelberg, 1999

Littmann, M., Willcox, K. & Espenak, F., *Totality: Eclipses of the Sun*, Oxford University Press, 1999

Lockyer, N., *Recent and Coming Eclipses*, 2nd edn, Macmillan & Co, London, 1900

Meeus, J., *Transits*, Willmann-Bell, Richmond, VA, 1989

Mitchell, S.A., *Eclipses of the Sun*, 4th edn, Columbia University Press, New York, 1935

Pasachoff, J. & Covington, M., *The Cambridge Eclipse Photography Guide*, Cambridge University Press, Cambridge, 1993

Peterson, I., *Newton's Clock: Chaos in the Solar System*, W.H. Freeman, New York, 1993

Ridpath, I. (ed.), *Oxford Dictionary of Astronomy*, Oxford University Press, Oxford, 1997

Stephenson, R., *Historical Eclipses and Earth's Rotation*, Cambridge University Press, Cambridge, 1997

Taton, R. & Wilson, C., *The General History of Astronomy*, Vol. 2A, Planetary Astronomy from the Renaissance to the rise of astrophysics, Cambridge University Press, 1989

Todd, M.L., *Total Eclipses of the Sun*, 2nd edn, Sampson Low, Marston Co., London, 1900

Zirker, J.B., *Total Eclipses of the Sun*, Van Nostrand Reinhold, New York, 1984

Photographic Credits

BnF = Bibliothèque nationale de France
BOP = Bibliothèque de l'Observatoire de Paris

p. 3: © Y. Delaye / Ciel & Espace
p. 4: © Roger Viollet
p. 6–7: © S. Brunier / Ciel & Espace

JOURNEYS OF AN ECLIPSE CHASER

p. 8–9: © S. Brunier / Ciel & Espace
p. 10: © S. Brunier / Ciel & Espace
p. 11: © S. Brunier / Ciel & Espace
p. 12: © S. Brunier / Ciel & Espace
p. 13: © S. Brunier / Ciel & Espace
p. 14: © S. Brunier / Ciel & Espace
p. 15: © S. Brunier / Ciel & Espace
p. 16–17: © S. Brunier / Ciel & Espace
p. 18: © S. Brunier / Ciel & Espace
p. 19: © S. Brunier / Ciel & Espace
p. 20–21: © S. Brunier / Ciel & Espace
p. 22: © S. Brunier / Ciel & Espace
p. 23: © S. Brunier / Ciel & Espace

THE STORY OF ECLIPSES, THE STORY OF PEOPLE

p. 24–25: *American expedition to Wyoming*, Total eclipse of the Sun of 29 May 1878 by E.L. Trouvelot, © BOP Meudon Collection
p. 26: © Werner Forman / AKG Paris
p. 27: © photo Jean–Loup Charmet
p. 28–29: © BnF, Department of Oriental Manuscripts
p. 30: © BnF, photo Jean–Loup Charmet
p. 31: © Bibliothèque des arts décoratifs, photo Jean–Loup Charmet,
p. 32: © BnF, Manuscript Department
p. 33: *top* History of China and India © Mary Evans / *Explorer*; bottom © photo Jean–Loup Charmet
p. 34: Costume of original inhabitants of British Isles, 1815, © Mary Evans / Explorer
p. 35: © Mary Evans / Explorer
p. 36: © BOP
p. 37: *top* © Musée du Louvre, RMN – H. Lewandowski; *bottom*, Total eclipse of the Sun, 29 July 1878, photo J. Bottet © Archives Larbor
p. 38–39: © BOP
p. 40: *top* © Musée du Louvre, RMN, photo Franck Raux; *bottom* BnF, Manuscript Department
p. 41: © BnF, Manuscript Department
p. 42: photo J. Bottet © Archives Larbor
p. 43: John de Holywood, known as Sacrobosco, *De sphaera mundi*, 1472 © BOP
p. 44: photo J. Bottet © Archives Larbor
p. 45: © Valenciennes, Musée des Beaux Arts, RMN, photo R.G. Ojeda
p. 46: Jean Mielot, *Avis collectif pour faire le passage d'outre mer*, © BnF, photo AKG, Paris
p. 47: *top* © photo Jean–Loup

Charmet; *bottom* © BnF, Reserve collection of rare books
p. 48: *top* © BOP; *bottom* © Royal Library, Copenhagen
p. 49: *top* © BnF, Print Department, Environment; *bottom* © BOP
p. 50: © BnF, Print Department
p. 51: © Observatoire de Paris Museum
p. 52: © BnF, Print Department
p. 53: © Keystone

SONGS OF THE ECLIPSES

p. 54: © O. Sauzereau / Ciel & Espace
p. 56: © BOP Bureau des longitudes
p. 57: H. Kraemer, *L'univers et l'humanité*, © photo Jean–Loup Charmet
p. 58: Rutherford et Proctor, *The Moon*, © BOP
p. 59: © BnF
p. 60: Private Collection, © photo Jose
p. 61: Richard Texier © ADAGP, Paris 1999

THE DANCE OF THE SUN AND MOON

p. 62: © M. Shirao /ABB / Ciel & Espace
p. 64: © NASA / Ciel & Espace
p. 65: © JPL / Ciel & Espace
p. 66: © BnF, département des estampes
p. 67: © BOP
p. 68: © BnF, Department of Maps and Plans
p. 69: Johann Gabriel Doppelmayer, *Atlas coelestis*, © BnF, Department of Maps and Plans
p. 70: © A. Fujii / Ciel & Espace
p. 72: top © BOP, bottom © C. Ichkanian / Ciel & Espace
p. 73: © BnF, Department of Maps and Plans
p. 74: © BnF, Department of Maps and Plans
p. 75: © BOP
p. 76: © BOP
p. 77: photo J. Bottet © Archives Larbor
p. 78: *top* Pierre Gassendi, *Opera Omnia, tomus quartus* © BOP; *bottom* © BnF, Manuscript Department
p. 79: Claude Buy de Mornas, *Atlas méthodique et élémentaire de géographie*, © BnF, Print Department
p. 80: © RMN
p. 81: © BnF, Department of Maps and Plans
p. 82: © BnF, Department of Maps and Plans
p. 83: © BnF, Department of Maps and Plans
p. 84: © Ciel & Espace

THE GREAT COSMIC CLOCKWORK

p. 86–87: © S. Brunier / Ciel & Espace
p. 88: © A. Fujii / Ciel & Espace
p. 89: © A. Fujii / Ciel & Espace
p. 90: © E. Renault, C. Amerge / Ciel & Espace
p. 91: © S. Brunier / Ciel & Espace
p. 92: © J. Dragesco / Ciel & Espace
p. 93: © J. Dragesco / Ciel & Espace
p. 94: © A. Fujii / Ciel & Espace
p. 95: *top* © C. Bemer / Ciel & Espace; *bottom* © A. Fujii / Ciel & Espace
p. 96: *top* © O. Staiger / Ciel & Espace; *bottom* © O. Staiger / Ciel & Espace
p. 97: © O. Staiger / Ciel & Espace
p. 98: © NASA / Ciel & Espace
p. 99: © E. Karkoschka / Ciel & Espace
p. 100–101: © NASA: Ciel & Espace
p. 100: bottom © NASA / Ciel & Espace
p. 101: bottom © NASA / Ciel & Espace
p. 102: © Denis Di Cicco, Leif Robinson, Sky and Telescope / Ciel & Espace
p. 103: © S. Brunier / Ciel & Espace
p. 104–105: © S. Binnewies, B. Schröter, P. Riepe, H. Tomsik / Ciel & Espace

BY THE LIGHT OF ECLIPSES

p. 106: © Ciel & Espace
p. 108: © S. Koutchmy, P. Martinez / IAP
p. 109: © J. Lodriguss, A. Gada / Ciel & Espace
p. 110: top © BOP, bottom © BnF, Reserve collection of rare books
p. 111: © APB / Ciel & Espace
p. 112: Aristarchus of Samos, *De magnitudinibus et distantiis solis et lunae*, © BnF, Reserve collection of rare books
p. 113: © Bibliothèque de l'Institut, photo Bulloz
p. 114: © Bianchetti/Selva
p. 115: © BOP
p. 116–117: *top* © BOP
p. 116: *bottom* © BOP
p. 117: *bottom* © BOP
p. 118: *top & bottom* © BOP
p. 119: *top* © BOP; bottom Royal Astronomical Society / Ciel & Espace
p. 120: *top* © F. Hammer, CFH; *bottom* © BOP
p. 121: *top* © NASA, STScI / Ciel & Espace; *bottom* © S. Brunier / Ciel & Espace

11 AUGUST 1999: THE LAST ECLIPSE OF THE MILLENNIUM

p. 122–123: © J.-P. Haigneré / CNES
p. 124: © D. Chung / Reuters/ MAXPPP
p. 125: © C. Soeters / Ciel & Espace
p. 126–127: © Baax
p. 126: *bottom* © G. Gios / Lp / MAXPPP
p. 127: bottom © L. Balogh / Reuters / MAXPPP
p. 128: © S. Chivet / Ciel & Espace
p. 129: *top* © E. Baudoin / Ciel & Espace, *bottom* © Mahfouz Abu Turk / Reuters / MAXPPP
p. 130–131: © O. Corsan / Lp / MAXPPP
p. 132–133: © C. et S. Brunier / Ciel & Espace
p. 132: *bottom* © I. Waldie / Reuters / MAXPPP
p. 133: *bottom* © J. Russell / Sygma
p. 134: © ABC Bassin Ajansi / Gamma
p. 135: © J.-L. Dolmaire / MAXPPP
p. 136–137: © E. Piednoel / Ciel & Espace

ATLAS OF ECLIPSES OF THE SUN AND MOON

p. 138–139: © S. Brunier / Ciel & Espace
p. 140: © S. Brunier / Ciel & Espace
p. 141: © S. Brunier / Ciel & Espace
p. 142–143: © S. Brunier / Ciel & Espace
p. 142: *bottom* © S. Brunier / Ciel & Espace
p. 143: *bottom* © S. Brunier / Ciel & Espace
p. 144: © S. Brunier / Ciel & Espace
p. 145: © S. Brunier / Ciel & Espace
p. 146: © S. Brunier / Ciel & Espace
p. 148–149: © A. Fujii / Ciel & Espace
p. 170: © A. Cirou / Ciel & Espace
p. 171: © S. Brunier / Ciel & Espace
p. 172–173: © M. Shiraro / APB / Ciel & Espace
p. 175: © S. Brunier / Ciel & Espace
p. 186: © Jonas / Kharbine Tapabor
p. 189: © Branger Viollet
p. 191: © BOP Bureau des longitudes
p. 192: © Grob / Kharbine Tapabor